Introduction to Interrater Agreement for Nominal Data

Roel Popping

Introduction to Interrater Agreement for Nominal Data

Springer

Roel Popping
Department of Sociology
University of Groningen
Groningen, The Netherlands

Additional material to this book can be downloaded from http://extras.springer.com.

ISBN 978-3-030-11670-5 ISBN 978-3-030-11671-2 (eBook)
https://doi.org/10.1007/978-3-030-11671-2

Library of Congress Control Number: 2018967726

Mathematics Subject Classification (2010): 62, 97, 60

This Springer imprint is published by the registered company Springer Nature Switzerland AG
The registered company address is: Gewerbestrasse 11, 6330 Cham, Switzerland

To Marika.

Preface

This book was born out of the feeling of need to help other researchers at times when I am not around. They must be able to set up, carry out, and report research themselves in which agreement between raters plays an important role. This concerns the extent to which there is consensus between independent raters in the classification of a sample of research units. It is not important here whether it concerns encoding texts or answers to open questions in surveys, observing in a behavioral study, diagnosing by an insurance physician, or determining PTSD by a psychiatrist. The book does not contain a detailed description of any of these types of research, but contains adequate material for a researcher to be able to build up such research.

The book contains the most important agreement measures we have and connects them with the most common research situations in which these measures are used. This should be sufficient for most researchers to start with. More important, the book tries to give backgrounds, due to which the researcher is expected to gain more understanding about establishing agreement between raters.

My thanks go to all those with whom I have collaborated in this kind of research in one way or another. They came to me with their problems. But I have learned a lot from those problems. This is reflected in the book.

Groningen, The Netherlands
December 2018

Roel Popping

Contents

Notation

In the text, the following symbols are used:

N	Number of units
m	Number of raters
c	Number of categories
r	Number of categories by the other rater in case this rater has an own set of categories
$f_{ij(gh)}$	Number of units in cell [i, j] of the agreement table for raters g and h. In case there are only two raters, this might be written as f_{ij}
$p_{ij(gh)}$	Proportion of units in cell [i, j] of the agreement table for raters g and h
$f_{(s)ij}$	Number of units in cell [i, j] of the agreement table for unit s based on the assignments by two raters
$f_{i(g)}$	Marginal total in row i of the agreement table, marginal total assigned by rater g to category i
$p_{i(g)}$	Marginal proportion in row i of the agreement table, marginal total assigned by rater g to category i
p_i	Proportion of units assigned to category i over all raters
$p_{i(mG)}$	Proportion of units assigned to category i over all raters from group G
$p_{i(mG+H)}$	Proportion of units assigned to category i over all raters from groups G and H together
w_{ij}	Weight in cell [i, j] of the matrix of weights
n_{si}	Number of times unit s has been assigned to category i by all raters or in all ratings
n_i	Number of times a unit has been assigned to category i by all raters or in all ratings
y_{sig}	Unit s has been coded in category i by rater g ($y = 1$) or not ($y = 0$)
v_{sii}	Number of identical codings assigned by two raters to unit s in the situation of multicoding

$v_{si(g)}$	Number of different codings used by rater g to classify unit s in the situation of multicoding
$f_{i_1 i_2 \ldots i_m}$	Number of agreeing assignments by all m raters to category i
$p_{i_1 i_2 \ldots i_m}$	Proportion of agreeing assignments by all m raters to category i

The symbol # is used to indicate marginal totals in tables.

Part I
Theory and Practice

Chapter 1
Introduction

When a physician examines a patient, it is desirable that the findings do not change when this patient is examined by a different physician. The patient will not feel confident if physicians seriously differ in opinion. Of course, this does not only apply to physicians, but applies to those who make judgments in general, especially in situations where it is impossible to establish the truth in an objective way.

In many fields of science, research units like persons (patients), events, texts, or broadcastings, observations are classified with respect to some characteristic into one of a limited set of categories, groups of units regarded as having particular shared characteristics. Such a classification is necessary because this is the only way to get a classification of the research units. A physician, for example, diagnoses patients or recognizes symptoms with respect to the type of disease, and a personnel officer classifies jobs with respect to the type of skills required. A rater in a text analysis study categorizes an article in a newspaper as mainly dealing with home or foreign news, but might also code open-ended interview data into analyzable terms. In observation studies, one looks at types of behavior that are exposed like, for example, the showing of solidarity or of antagonism. Even astronomers might classify star formations as a spiral, beam, or one of the other types that have been distinguished among these formations. In specific investigations, pictures, sounds, and gestures might be judged. Dozens of examples from many fields could be added. In many situations, the measurement involves at least some ambiguity in the characteristics of interest in the unit. The measurement task often involves a subjective moment when the decision is to be made which category applies most for a specific unit. The person, rater, responsible for the coding is supposed to know how to deal with this ambiguity. This rater is an expert or had a training in how to perform the coding. The only way to find out whether the decision by the rater can be trusted is by having the classification also done by another rater, who is equally skilled as the first one and who operates independently from that first one. The scores should not represent the rater's personnel opinion, but should be in line with the view that is common in the culture in which the study is performed. The more identical the assignments per unit, the more trust an investigator can have in the final classifications. This situation is often found in the fields of the social,

R. Popping, *Introduction to Interrater Agreement for Nominal Data*,
https://doi.org/10.1007/978-3-030-11671-2_1

behavioral, political and communication sciences, marketing, biology, and medicine. The variable that is constructed in this way will be confronted with other variables in the study.

In any scientific research, associations between different may be blurred by errors in the assessment of characteristics. The possibility exists that important associations are overlooked. Associations that are discovered may be hard to interpret because different studies may result in different quantitative conclusions. The specification of disagreement may be a first step to a better agreement among raters: Which disagreements do frequently occur and which raters often have different opinions. This may lead to a further standardization of the way of judging.

Interrater agreement is the widely used term for the extent to which independent raters evaluate a characteristic of a message or artifact and reach the same conclusion. The classifications that are made always concern one specific characteristic. The importance of interrater agreement lies in the fact that it represents the extent to which the data collected in the study are correct representations of the variables measured. This book deals with the question of how to compute interrater agreement. More precisely, it will be on agreement on data measured at a nominal level of measurement. At this level, one simply names or categorizes units. As an example, in case we would look at the variable 'eye color,' the categories might be 'blue,' 'green,' 'brown,' and 'hazel.' In text analysis studies, one has to think about whether a sentence in a text is about politics, economics, or none of these, as in the earlier example. In observation studies, one looks at whether behavior that is shown expresses solidarity, antagonism, or something else. A psychiatrist classifies patients as schizophrenics, manic depressives, psychotic depressives, or in some other psychotic group. The essential point about nominal scales is that they do not imply any ordering among the categories. A nominal scale consists of a (limited) number of mutually exclusive and exhaustive categories. Higher levels of measurement are the ordinal level, where the categories also are ordered, and the interval level, which is one more step higher. The interval level of measurement not only classifies and orders the measurements, but it also specifies that the distances between any two succeeding categories is the same. A consequence of the restriction to data on the nominal level is that agreement statistics based on Pearson product moment correlations or upon analysis of variance techniques (which assume interval categories) are ruled out.

If the scores assigned to units by two or more raters are (almost) the same based on some criterion, the coding task is said to have a good reliability. This indicates most of all that there seems to be little measurement error and that the codings are reproducible. Some researchers, as we will come to see, go further. They add that the raters basically are exchangeable and that such an exchange does not affect the coding process. Having reliable data means the investigator has generated data in his or her study that can be relied upon and therefore can be studied, theorized, or used in pursuit of practical decisions or further analysis.

Table 1.1 contains the assignments by two raters in a text analysis study using the theoretical premise that people motivate themselves and others by persuading them that their actions are possible, impossible, inevitable, or contingent. This persuasion is accomplished using reality claims (Popping and Roberts 2009).

Table 1.1 Two raters' assignments of four reality claims

		Rater 2					
		Possible	Impossible	Inevitable	Contingent	Not coded	#
Rater 1	Possible	217	24	0	0	15	256
	Impossible	2	120	0	0	2	124
	Inevitable	0	0	84	0	2	86
	Contingent	0	1	0	8	0	9
	Not coded	4	3	3	2	0	12
	#	223	148	87	10	19	487

Sentences containing such claims had to be coded into one of these four categories. At first glance, the categories seem to be difficult to understand. After a careful explanation however, the rater can recognize them very well. Abstract categories like these are often used when judgments are made. It is possible that one rater noted a claim while the other did not. Therefore, an additional category—not coded—was necessary. An example of a sentence to be coded is the following:

> An ethical point of view is usually justified in case of all human actions, because its subjects, humans *cannot* be separated from their actions (Népszabadság, April 1, 1994, p. 14).

This sentence does not contain a reality claim that motivates a person; the word 'cannot' as used here does not refer to an intention. Another example is the sentence:

> Hungarian society is just learning the rules of democracy, and developing the rules of the game, which *must* be adopted by politicians (Népszabadság, October 15, 1993, p. 15).

This sentence is encoded as a reality claim, even though its grammatical subject is not a person. This is because its use of 'must' as the reality claim *inevitable* is in passive voice and has people (namely politicians) as its semantic subject. Coded is the rater's interpretation of the sentence, usually it is not known which choice is the correct one, but, if there are clear rules to follow, raters should come to the same assignment. The raters had the same training. They are equally skilled, and therefore, they should be interchangeable. It is during the coding process that data are generated for the investigation. Only after data have been generated, the investigator is able to link each individual datum and the whole data set to the units of interest.

The comparison of all assignments on which the raters agreed with respect to the type of reality claim results in the score on an agreement index, which is a descriptive index. The simplest index, but not the best as will be explained, consists of the proportion of units on which both raters agree. This is the proportion of units on the diagonal of the table showing the assignments by the raters; for the data in Table 1.1, the proportion of observed agreement, P_o, is (217 + 120 + 84 + 8 + 0)/ 487 = 0.88. Disagreement is found with respect to the units that are not on the diagonal of the table.

The difference between reliability and agreement is described very well: 'Interrater reliability provides an indication of the extent to which the variance in the ratings is attributable to differences among the rated subject. … Interrater agreement represents the extent to which the different judges tend to assign exactly the same rating to each object' (Tinsley and Weiss 2000: 98). In the above study, the question was whether the codes are a good reflection of reality, that is, to say whether they are valid. If the degree of agreement is higher than a certain criterion, it is stated that the codes are reliable. This is an important step toward validity. By relating the variable with these codes as scores to variables that relate to something else from the content of the sentences from which they are taken, for example, their rationale (politics, economics, and so on) and at the time (year) on which they were published, it is possible to get view on developments in a culture regarding the direction in which that culture should change according to the residents. Most investigators connect agreement to reliability as is done here.

The example presented above concerns data at the nominal level of measurement. In practice, the rater sometimes needs two steps to come to a decision. First, the rater must decide whether the attribute is at present or not. If it is at present the second decision is how it shows itself. The first decision with respect to the example used is whether a reality claim is found in the sentence under investigation or not. If so, the next decision concerns the type of reality claim. The two decisions are always reported in one table and one outcome of the index is reported. But actually, there are two sources of disagreement. Raters may differ in the *definition of the attribute itself* (does the sentence contain a reality claim: yes or no); or in their definitions of *specific categories* (which type of reality claim is at present in the sentence).

The number of different interrater agreement indices found in the literature is remarkably high. This is partly due to the discovery that the measurement of agreement is not only relevant in the simple situation in which two raters have each classified a sample of N units ($N > 1$) into the same set of c pre-specified categories. Many variations, extensions, and refinements of this simple situation are met in empirical research, and moreover, the ways and goals of measuring agreement exhibit a similar variation. Such a specific setting is called here an 'empirical situation.' In this text, many of these situations will be discussed. In practice, we frequently see that (a variant of) a measure is used in a situation where it should not be used. As we will see later, this often concerns the kappa-statistic, a measure which is very popular. This book will not give a complete overview of the literature and will also not treat all specific research situations that might be distinguished; for developments up to 1990, see Popping (1992). Since books have appeared covering (almost) all at that moment available measures, for example: Gwet (2014), Shoukri. (2011).

The idea in this book text is to help the investigator, who is to use interrater agreement indices, get started. We start from a sample of units that is at least rated twice.

Agreement for data at an ordinal or interval level of measurement is not considered other than that it is mentioned that these levels can be approached by using weights. Usually, the reliability for data at these levels is computed in another way.

However, one can use weights in the formulas to cover these levels. Weights will have to be entered into the indices for these data. The issue will come back shortly at the end of Chap. 3.

Studying agreement is not useful in qualitative research. The concept of reliability is considerably overrated in this type of research. Here, coders often notice different things in the data or interpret them differently. If they do, then these issues need to be resolved. The idea of 'saturation' is relevant here. Campbell et al. (2013) have tried to develop a method for computing agreement in a qualitative study. They found that especially the way transcripts are unitized biases coding and inflates agreement scores and therefore interrater reliability. However, one must look carefully at what is meant by qualitative research. It occurs often that studies deal with 'qualitative research,' but that these are actually descriptive (quantitative) studies. This can be a study where the investigator is still looking for the best set of categories to be used. It is possible that units are available, and a category is assigned to each unit. However, the categories belong to a set that is developed by the rater, and each rater uses an own set. Here too, interrater agreement can be computed (Popping 1983a). Afterward one can see whether and where these sets of categories differ and, if so, find out what the rater has been looking for. This might help in defining the set of categories to be used in the actual study. In a real qualitative study however, where the goal is to construct theories, one does not work with well-defined units.

The judgments of only one unit by many raters are not a problem of interrater agreement. Such data one gets when one asks lots of people (by preference a sample) to judge, for example, a painting on some (one-dimensional) scale. The score denotes an average appreciation for the unit. For such, the reader might start with Lindell and Brandt (1997).

Also, a note on terminology is necessary. The name 'rater' denotes the agent that produces a classification (although in some applications this is not one human being). In many texts, the words 'coder,' 'judge,' 'observer,' 'annotator,' or 'classifier' are used, sometimes even the description 'diagnostic test.' In general, 'units' is used for that what is classified (although these may be human beings and not things as in our example of patients). In literature at least also, the words 'respondent,' 'person,' 'patient,' 'observation,' 'subject,' 'object,' 'phenomenon,' and 'unit of analysis' are found. To make it more complex, often not the complete unit is investigated, but only an attribute of the unit, i.e., a quality or characteristic that is a particular part of a unit. Where possible I will try to use the word 'unit,' but at some places 'person' or 'observation' will be used as otherwise the text sounds curious.

One might distinguish between 'sampling units' and 'recording units.' The first are the entities that are sampled and that are usually used in the statistical analysis. The second refers to the part that is recorded. Sometimes the two coincide, but often a sampling unit contains several recording units. In the study from which Table 1.1 was taken, the sampling units are newspaper editorials that were sampled and the recording units were all sentences in these editorials that contain a reality claim. Some editorials do not contain such a claim at all; others contain several claims. In observation studies, the test person is generally the sampling unit; the occurrence of the specific type of behavior that is under study commonly is the recording unit. Usually, the total observation time is divided into smaller time intervals and it is investigated whether the behavior of interest occurs in each interval or not. For agreement, only the recording units are relevant.

First reliability will be explained to the reader, as this usually is the main reason to compute agreement. Here, it is also important to pay attention to validity. In any empirical study, reliability of the data is an essential part of the validity of that study. With this context information in mind, agreement can be explained. In a long chapter, the reader will get an explanation of what agreement is and will hear about the role of the units that are investigated, the number of categories that is used and the number of raters to be used, but also about their qualifications. Some agreement will be based solely on chance. A correction for such agreement is necessary. Investigators have no doubt on this, but how to define chance expected agreement seems an unsolved question. Two issues are important. First, nearly every investigator takes the position that the distribution of units over categories should be used, but questionable is: Which distribution? Second, in case the distribution, that is finally used, is much skewed the estimate of the agreement index will collapse. What to do now? Also, attention is given to the question of how the outcomes an agreement index in general should be understood. The most common situations in the field of research are discussed. Training of raters and problems in one of the most complex types of study, the observational one, get attention.

In the next chapter, indices are presented. No advanced mathematical explanations will be given, neither will all indices be presented that have been proposed once, and a lot of discussions that did not result in something that still persists are not mentioned. The focus is on nominal data and dichotomous data. Higher levels of measurement get very little attention; these levels can be approached by introducing weights. This implies that there is also no treatment of correlation indices, especially the intraclass correlation coefficient or of variance analytical designs, which are common for especially interval data.

References

J.L. Campbell, C. Quincy, J. Osserman, O.K. Pedersen, Coding in-depth semi-structured interviews: problems of unitization and inter-coder reliability and agreement. Sociol. Methods Res. **42**(3), 294–320 (2013)

K.L. Gwet, *Handbook of Inter-Rater Reliability. The Definitive Guide to Measuring the Extent of Agreement among Raters* (Advanced Analytics LLC, Gaithersburg, USA, 2014)

M.K. Lindell, C.J. Brandt, Measuring interrater agreement for ratings of a single target. Appl. Psychol. Meas. **21**(3), 271–278 (1997)

R. Popping, Traces of agreement. On the dot–product as a coefficient of agreement. Qual. Quant. **17**(1), 1–18 (1983)

R. Popping, C.W. Roberts, Coding issues in semantic text analysis. Field Methods **21**(3), 244–264 (2009)

R. Popping, *Taxonomy on Nominal Scale Agreement* (iec *Pro*Gamma, Groningen, 1992)
M.M. Shoukri, *Measures of Interobserver Agreement and Reliability* (CRC Press, Boca Raton, Fl, 2011)
H.E.A. Tinsley, D.J. Weiss, Interrater reliability and agreement, in *Handbook of Applied Multivariate Statistics and Mathematical Modelling*, ed. by H.E.A. Tinsley, S.D. Brown (Academic Press, San Diego, 2000), pp. 95–124

Chapter 2
Reliability and Validity

The amount of agreement that is computed after a coding task has been finished is often used as an indication for the reliability of this coding. This refers to the consistency of the coding. The reliability in its turn is among others used to say something about the validity of the data, about whether the data measure what they are supposed to measure.

In this short chapter, some necessary remarks are made with respect to reliability and validity. After that, we can switch to agreement.

2.1 Reliability

In statistics, reliability refers to the consistency of a research study or measure. A measure having a high reliability produces similar results under consistent conditions. The term reliability as used with respect to coding refers to rater consistency and not, as commonly, to data consistency. A situation of data consistency, for example, is looked for when a research is replicated. If the findings of the replication are consistent with the original findings, we conclude the data are reliable. A correlation coefficient can be used to assess the degree of reliability for data at a higher level of measurement. In the situation of nominal data, the χ^2-test is used. If a test is reliable, it should show a high positive correlation. The purpose of reliability assessment in the context of this book is to assure that a data generating process can be replicated elsewhere, by other investigators, using the same coding instructions and the same units but different raters, and can therefore be relied upon in the present study. Agreement refers to the extent to which raters agree in their coding of units. Very often, this concerns a characteristic of the unit or behavior as exposed by the unit. Coding is an analytical process in which data are categorized with respect to a specific characteristic to facilitate analysis. A coding scheme is used that indicates the available choices, but that also tells how to handle in complex situations. This categorization is preceded by an interpretation process.

R. Popping, *Introduction to Interrater Agreement for Nominal Data*,
https://doi.org/10.1007/978-3-030-11671-2_2

Usually, a coding task is performed by humans, called raters. If different raters perform the same coding task, one hopes they end up with identical assignments; that is, all raters code all separate units into the same category. When this occurs, we have rater consistency. According to Holsti (1969: 135), who is in the field of text analysis: 'Reliability is a function of coders' skill, insight, and experience, clarity of categories and coding rules which guide their use and the degree of ambiguity in the data.' The importance is read from the following: 'interjudge reliability is often perceived as the standard measure of research quality. High levels of disagreement among judges suggest weaknesses in research methods, including the possibility of poor operational definitions, categories, and judge training' (Kolbe and Burnett 1991: 248). It might also denote a lack of diagnostic accuracy (Uebersax and Grove 1990). A high level of agreement between raters, however, does not guarantee an individually correct coding. Interrater agreement indices indicate the extent to which the assignments by the raters are reproducible.

To assess the degree of reliability, an agreement index is needed. Such an index numerically describes the extent of replicability of the nominal measurement. In many applications, the classifications are made by raters; therefore, the term interrater reliability is often used, but applications are not confined to this situation. Agreement is what we compute; reliability is what we infer from it.

Agreement is very often considered as a special kind of association. There are differences, however. It is important to determine the similarity of the content of behavior (in a broad sense) between raters in general with the degree of identity of this behavior. The behavior of one rater does not have to be predicted from that of the other rater. In the case of association, one investigates the strength of the linear relationship between variables. Here, the goal is to predict the values of one variable from those of the other variable. With regard to agreement, most important is the similarity of the content of behavior between raters, with the goal of determining the degree of identity of this behavior [see Popping (1988), about the same reasoning is followed by Krippendorff (1987)]. One looks at agreements or disagreements among multiple descriptions or manifestations generated by a coding procedure, regardless of who enacts that procedure. The basic idea of an agreement index is looking at the fraction of units on which raters agree in assigning to a category. The difference between agreement and association is shown in Table 2.1. The data in table A at the left show both perfect association and perfect agreement. For perfect agreement, it is necessary that the raters used the same set of categories, here categories *a*, *b*, and *c*. The data in table B at the right only show perfect association, there is no agreement at all. For this reason, it is not necessary that both raters used the same set of categories. Perfect agreement implies perfect association, but perfect association does not imply perfect agreement.

Similarity is different from association in that it usually focusses on the comparison of pairs of units while association is usually between variables. In the situation of similarity, often yes-no data are involved. Therefore, similarity is even further away from agreement than association. The main difference with distance is that there also triangle inequality is required, and the sum of the lengths of any two sides must be greater than or equal to the length of the remaining side.

Table 2.1 Tables with perfect association, with resp. without agreement

		Rater 2					Rater 2			
	A	*a*	*b*	*c*	#	B	*a*	*b*	C	#
Rater 1	*a*	30	0	0	30	*a'*	0	30	0	30
	b	0	30	0	30	*b'*	0	0	30	30
	c	0	0	30	30	*c'*	30	0	0	30
	#	30	30	30	90	#	30	30	30	90

High agreement is generally accepted as evidence that raters have coded units (observed behaviors) similarly and consistently. If this is the case, then the assumption is made that variations in the scores provided by one rater can reasonably be attributed to actual variation in the unit (in observational studies, this would be the unit's behavior), rather than to variation or inconsistency on the part of the raters. There is a strong parallel between agreement and the precise psychometric definition of reliability: An attempt is made to separate the unit's performance from errors introduced into the recorded data while measuring that performance (by rater inconsistency or random measurement error). With a high agreement index, eventually in addition to the conclusion that the raters are interchangeable, an investigator may also wish to infer that data from any one of the raters are relatively free from measurement error. Codings are said to be reliable when the outcome of the agreement index is above some criterion decided upon by the investigator. This criterion is discussed in more detail in Chap. 3.4.

The term reliability points to three different issues: stability, reproducibility, and accuracy. When a text analysis is performed by using some computer program, and the dictionaries, the trivial words list, and the lists of negative words are not changed, the second analysis will come up with the same results as the first one. This refers to stability; in a next analysis, the same results are found, or better, one will get the same results when repeated measurements are made using similar instruments and under similar conditions. In the situation where the coding is performed by some human rater, one also hopes to find no real differences between the first and the second coding. For example, the reliability checks Bara (1987) mentioned in her study are based on the second coding of documents by the same rater. She did not report the results, and this does not help the reader. Here, intrarater agreement is considered. The assessment of this type of agreement is problematic, especially because of statistics involved. If it is possible to repeat the coding process, for example because the data are videotaped, the second assignment is not independent of the first one. Too often the rater will remember how the coding was done the first time and why it was done in this way. This is also true when a text analysis study is performed. But in case there are no data that can be used again and therefore the research must be repeated completely, it will almost become impossible. One would have to replicate identical behaviors and observational conditions.

When coding is performed by humans, there are usually several raters. It should not matter now which rater does which coding; all raters come up with the same results.

This holds under the conditions that the categories and the coding instructions are clear, and that the raters have been trained. In case small differences are found, these are based on chance. Indices are available to test whether there are really differences between raters, or whether categories are hard to distinguish. In case there are really differences, this is often due to ambiguous or inconsistent coding rules, or to cognitive factors like abstraction mismatches. In case specific categories are often exchanged, it might be that these categories are hard to distinguish, or not exclusive. The reliability that is considered here is known as reproducibility. In the situation of reproducibility, it is possible that two independent raters score units inaccurately, but still show high agreement: They may both be inaccurate in the same way.

Accuracy is investigated when a rater is compared to some established criterion, usually denoted as 'standard' or even 'gold standard.' It indicates whether the measurement is a true representation of what is being observed. This assigning often occurred previously and was done by an expert. Experts' scores are often used in a training situation. In a medical training situation, for example, students have to diagnose patients with respect to some problem. Their diagnoses are compared to those by a senior physician, who is the expert. It is assumed here that the diagnoses or decisions by the physician are the correct ones.

In the situation of reproducibility, it is allowed that raters agree on an assignment; nevertheless, the assignment is not correct. In the situation of accuracy, it is known how the assignment should be. The correct ones are known thanks to the standard. Now, deviation from a norm can be investigated. Sometimes a slightly different definition of accuracy is used. In this situation, it is meant that the rater codes as if he or she were an expert. To give an example: Someone very familiar with a specific field will be more accurate in encoding utterances from that field than a novice rater with less expertise. Therefore, this does not have to be problematic in training situations. In 'real' situations, it can be dangerous, even an 'expert' can make mistakes.

For latent content, i.e., concepts that are not measured directly but indirectly by using an indicator, the raters must provide subjective interpretations based on their own mental schema and this 'only increases the importance of making the case that the judgments of coders are intersubjective, that is, those judgments, while subjectively derived, are shared across coders, and the meaning therefore is also likely to reach out to readers of the research' (Potter and Levine-Donnerstein 1999: 266). The opposite is manifest content, this refers to the concepts that are physically present and countable. A high level of agreement is a necessary step in validating the coding task, although not a sufficient one.

One needs to have clear arguments for computing rater agreement. These arguments are relevant, because they help determining which statistical method is best. Besides, one needs a theory for analyzing agreement. In the next chapter, it will be explained that one can look at agreement in different ways. Later will come that nearly all statistical methods for analyzing agreement have their own assumptions. If one has not thought about the data from a theoretical point of view, it will be hard to select an appropriate method. In a text analysis study, all raters are equal (after having had training), and as indicated before, usually there is not a correct answer. The raters usually have to code whether an attribute is there or have to indicate a specific behavior. Here, reproducibility is measured.

Some authors (e.g., Herbert and Attridge 1975) claim that observer agreement is not a measure of reliability at all. In their view, observer agreement does not by itself assess observer accuracy unless it is compared with some previously established standard. Investigators in the field of medicine might claim the same condition. Hollenbeck (1978) adds that observer agreement also does not assess stability unless it is repeated over repeated trials.

Sometimes agreement studies are an attempt to *validate* a new rating instrument. Such an instrument usually consists of several items. One wants to know whether the items are understood as they are intended and whether comparable raters assign the answers in an identical way. A good result is found when the agreement per item is high. If it is not high, there might be problems with the formulation or in the assigning. When the items seem correct, one might want to combine them in some scale, so that per unit one score for the characteristic measured in the instrument results. This way of working is important when there is no criterion variable.

An example of such a study is reported in Spanjer et al. (2009). The authors investigated whether the provision of detailed information on participation and activity limitations, compared with medical information alone, influences the assessment of work limitations (the items) by physicians. Three groups each of nine insurance physicians used written interview reports to assess work limitations in 30 patients with low back pain or lower extremity problems.

Each group was given different kinds of information on the patient: The first group received only medical information; the second group received detailed information on participation and activity limitations; and the third group was provided with both types of information. Agreement percentages within the groups and differences between the groups in scores given on the work limitation items in the Functional Ability List were measured. The groups showed no important differences in agreement percentages (mean percentage approximately 80%). The physicians who received either medical information or both forms of information indicated fewer work limitations compared to physicians using detailed information on participation and activity limitations. Information on participation and activity limitations provided by the patient has only limited influence on interrater agreement. However, there was a significant difference in scores on assessed work limitation items compared to medical history-taking alone. Therefore, in disability assessment interviews, physicians should ask for medical information as well as detailed information on participation and activity limitations.

Often the agreement is determined based on the coding of a representative part of the available data. If this agreement turns out to be sufficient according to the investigator, it does not matter anymore who did the coding. Therefore, the coding of the remaining data is divided between the raters participating in the agreement study.

A reliability check is not always performed. Lombard et al. (2002) found that only 69% (N = 137) of the articles they investigated in which a content analysis was reported contained any information on interrater reliability. In these articles, even few details were provided in the average 4.5 sentences devoted to reliability. Important information was ambiguously reported, not reported, or represented inappropriate decisions by researchers. Riffe and Freitag (1997) report that in 56% of the publications using content analysis in the journal Journalism Quarterly in the years 1971–1995, interrater reliability has been computed. Lacy and Riffe (1996) found that in many text analyses, only an arbitrary sample of all coded units is rated by multiple raters. This subsample can be considered as drawn from a population of units, which is actually the original sample. Lacy and Riffe are referring here to the units (sometimes sampling units), but a sample of recording units is also possible. Not the text as a unit is coded, but only the distinct units that coders are instructed to categorize, the recording units, in the text. This sample needs to have reliability estimates representing the real population. Therefore, the size of the subsample must permit a known degree of confidence that the agreement in this subsample is representative of the pattern that would occur if all units are coded by all raters. This is why investigators need to estimate the size for the subsample required for valid reliability assessments.

Sometimes people confuse intraclass correlation (ICC) indices with agreement indices. The ICC indices deal with the question what proportion of the variation in rating is due to rate performance rather than rater error. This comes back later.

2.2 Validity

Validity deals with the question whether the findings in the investigation represent real events. In a text analysis study, the investigator is especially confronted with the issue when the categories and search entries are developed and applied. In the development phase, the main issue is first of all whether the categories that are defined are the relevant and realistic ones for this research question. Next comes the question whether the correct method of text analysis is applied. The question about the categories can also be posed in the coding phase. In this phase, the question is to be added whether the search entries are used in a correct way, do they really stand for the categories as intended.

Reliability does not automatically imply validity. A reliable measure is measuring something consistently. This does not necessarily mean that one measures what one wants to measure. Reliability places a limit on the overall validity of a test. A test that is not perfectly reliable can never be perfectly valid. The relation between reliability and validity is explained by Campbell and Fiske (1959: 83): 'Reliability is the agreement between two efforts to measure the same attribute through maximally similar methods. Validity is represented in the agreement between two attempts to measure the same attribute through maximally different methods.' One must be careful, however; a high reliability does not mean there is validity. It is a necessary condition, but not a sufficient one. On the other hand, a low reliability implies that there are questions with regard to the validity.

Three different interpretations exist of the requirement that the results of a measurement should be about 'what it is intended to measure.' These are based on different meanings of 'realism' (Andrén 1981: 51):

- The realism of a certain set of data consists of its *correspondence* to some facts, i.e., its truth;
- The realism of a certain set of data consists of its *connection* with some significant problem or with the purpose of the study, i.e., its relevancy;
- The realism of a certain set of data consists of its *correspondence* with precisely those facts that are *connected* with some real problem or the purpose of the study, i.e., truth and relevancy.

The meaning of the term 'validity' often oscillates between the choices (2) and (3). The first choice refers to reliability, in particular, to the view of accuracy.

Face validity or content validity is a means that is generally used first, but often this is not sufficient. It refers to the way things are considered at first glance. It is estimated by judging how well the instrument represents a whole class of situations, knowledge domains, or skills about which conclusions are drawn. In the development phase of the instrumental investigation, one constantly has to wonder: 'are these the relevant categories and search entries?' in the coding phase the question is: 'are these the correct search entries and are they used in a correct way?' Face validity gives an impression, but just that.

In many studies, the investigator wants to generalize the results that were found. In that case, the validity is to be claimed for a larger class of characteristics than just the raters. Cone (1977: 414 ff) depicted the characteristics in which an investigator performing an observational study is most interested. Rephrased for a text analysis or observational study, these characteristics are:

1. The *rater*, it should not matter who performed the coding or did the observation, see before. The rater is trained;
2. The *variable,* does the variable measure what it is supposed to measure (this can be investigated by considering the internal consistency or by correlating the variable to another variable measuring the same characteristic although measured in another way);
3. The *time*, this has to do with the extent to which data based on text or on observation dating from one point of time are representative of data from texts written or observations registered at other points of time;
4. The *setting*, here the question is whether data collected in one setting are representative for those obtainable in another setting;
5. The *method*, here this would mean whether data collected in a text analysis or observational study are comparable to data collected in another way; and
6. The *medium*, this concerns the comparability of data based on two or more different media. This refers especially to speech versus text. I would also consider text versus text, for texts from different media might vary in length and level of complexity. Observations will usually be based on video. There are several ways to broadcast a video. The camera (and the person behind it) can be visible or not.

These aspects come back in at least one type of external validity that follows.

The external validity measures the correspondence between variations inside the process of analysis and variation outside that process, whereas internal validity assesses the consistency of results across items within a test. Moreover, external validity indicates whether findings represent the real units in the context of the data as claimed. This includes many aspects. Krippendorff (1980: 158) distinguishes three aspects for which validity is to be considered. The first aspect, including two types of validity, is related to the data: semantic validity and sampling validity; the second, also including two types, is related to pragmatics (the subfield that studies the ways in which context contributes to meaning): correlational validity and predictive validity; and the third one is process oriented: construct validity.

Semantic validity. In text analysis, a meaning category is said to have semantic validity if there is consensus among persons familiar with the language and texts under study that the words or phrases rated into this category do reflect the category's meaning (Krippendorff 1980: 159–62). Assurance of semantic validity requires more than a face validity check for similarities among words' meanings when taken out of context. It is for semantic validity that, for example, raters performing text analysis use software to verify that words are used in ways that are appropriate to their respective meaning categories. Overcoming all types of ambiguity in texts, idiomatic, illocutionary, and relevance ambiguity falls under the semantic validity. In other fields, this type of validity deals with the question whether the meaning categories really reflect what they are supposed to reflect, this refers most of all to the behaviors that are included and that are not included. This type of validity is necessary to secure that results of studies are realistic in the second sense as indicated by Andrén. The aspects mentioned by Cone that get most attention concern the rater and the variable. Reliability usually is seen as part of semantic validity, and it tells how well that what should be measured is measured.

The coding process in a study also fits under the semantic validity. Therefore, the determination of the reliability of this coding and consequently also the computation of agreement fall under this type of validity. Interrater reliability tells how well the coding task has been performed, and it does not say that units have been coded into the correct categories. For such, it is necessary that the codings are compared to the codings by a standard, who (by definition) assigned the correct category to the units.

Sampling validity. Assesses the degree to which available data are either an unbiased sample for a universe of interest or sufficiently similar to another sample from the same sample so that the data can be taken as statistically representative of that universe. This also fits in the second sense as indicated by Andrén. Remarks on sampling earlier in this chapter are connected to this type of validity. This type of validity precedes for the greater part the aspects mentioned by Cone; time, setting, and eventually dimension, however, can be relevant.

Correlational validity. The degree to which findings obtained by one method correlate with findings obtained by another and thus justify their substitutability. This relates to the method as mentioned by Cone. This type of validity also fits in the second sense according to Andrén.

Predictive validity. This type of validity refers to the ability of an instrument to predict events for which evidence can only be found at a later point of time. Carley (1986, 1988) made cognitive maps of her respondents who were students, based on these maps she predicted how the students would vote when the new tutor for their dormitory is to be elected. After the elections, she could verify whether her predictions were correct. The results can be used in a prediction, but it is required that this prediction holds. This fits in the third sense as indicated by Andrén, and both truth and relevancy are required. The main aspects according to Cone that deserve attention are time and setting.

Closely related to predictive validity is *concurrent validity*. Here, it is rated how much the results correspond to criterion data that are available at the same time.

Construct validity. Investigates the qualities measured by a test, it can be decided upon after all inferences have been performed. This type of validity 'is not only concerned with validating the measure, but also the theory underlying the measure' (Holsti 1969: 148). So, the validity is on the results of an investigation and the theory in which the problems behind the investigation are represented. This again seems to fit in the second sense as indicated by Andrén.

Andrén (1981: 54) proposes to identify the reliability of raw data with their truth and the validity of raw data with their relevancy. The truth deals with the first sense of realism. This is actually the reliability. Relevancy deals with the second sense of realism, this is the validity. When these are combined, they cover the third sense of realism.

References

G. Andrén, Reliability and content analysis, in *Advances in Content Analysis*, ed. by K.E. Rosengren (Sage, Beverly Hills, 1981), pp. 43–67

J. Bara, Israel 1949–1981, in *Ideology, Strategy and Party Change: Spatial Analyses of Post-war Election Programmes in 19 Democracies*, ed. by I. Budge, D. Robertson, D.J. Hearl (Cambridge University Press, Cambridge, 1987), pp. 111–133

D.T. Campbell, D.W. Fiske, Convergent and discriminant validation by the multitrait–multimethod matrix. Psychol. Bull. **56**(1), 81–105 (1959)

K. Carley, An approach for relating social structure to cognitive structure. J. Math. Sociol. **12**(2), 137–189 (1986)

K. Carley, Formalizing the expert's knowledge. Sociol. Methods Res. **17**(2), 165–232 (1988)

J.D. Cone, The relevance of reliability and validity for behavioral assessment. Behav. Ther. **8**(3), 411–427 (1977)

J. Herbert, C. Attridge, A guide for developers and users of observation systems and manuals. Am. Educ. Res. J. **12**(1), 1–20 (1975)

A.R. Hollenbeck, Problems of reliability in observational research, in *Observing Behavior*, vol. 2, ed. by G.P. Sacker (University Park Press, London, 1978), pp. 79–98

O.R. Holsti, *Content Analysis for the Social Sciences and Humanities* (Addison Wesley, London, 1969)

R.H. Kolbe, M.S. Burnett, Content-analysis research: An examination of applications with directives for improving research reliability and objectivity. J. Consum. Res. **18**(2), 243–250 (1991)

K. Krippendorff, *Content Analysis: An Introduction to Its Methodology* (Sage, Beverly Hills, CA, 1980)

K. Krippendorff, Association, agreement, and equity. Qual. Quant. **21**(1), 109–123 (1987)

S. Lacy, D. Riffe, Sampling error and selecting intercoder reliability samples for nominal content categories. Journal. Mass Commun. Q. **73**(4), 963–973 (1996)

M. Lombard, J. Snyder-Duch, C.C. Bracken, Content analysis in mass communication: assessment and reporting of intercoder reliability. Hum. Commun. Res. **28**(4), 587–604 (2002)

R. Popping, On agreement indices for nominal data, in *Sociometric Research*, vol. I, ed. by W.E. Saris, I.N. Gallhofer (McMillan, London, 1988), pp. 90–105

W.J. Potter, D. Levine-Donnerstein, Rethinking validity and reliability in content analysis. J. Appl. Commun. Res. **27**(3), 258–284 (1999)

D. Riffe, A.A. Freitag, A content analysis of content analyses: twenty-five years of journalism quarterly. Journal. Mass Commun. Q. **74**(4), 873–882 (1997)

J. Spanjer, B. Krol, R. Popping, J.W. Groothoff, S. Brouwer, Disability assessment interview: the role of concrete and detailed information on functioning besides medical history taking. J. Rehabil. Med. **41**(4), 267–272 (2009)

J.S. Uebersax, W.M. Grove, Latent class analysis of diagnostic agreement. Stat. Med. **9**(5), 559–572 (1990)

Chapter 3
Interrater Agreement

As already indicated, interrater agreement refers to the extent to which raters agree in their classification of a sample of units. The units are independent from each other. This implies that all units in the population of interest must have the same probability of inclusion in the sample. In the ideal case, the sample contains a perfect representation of the population. In case an investigator wants to compute such an agreement, there should be a reason to do so. One wants to know, for example, about the reproducibility of the coding task, interchangeability of raters, the quality of an additional rater, complexity of coding, recognizability of different categories, contribution to validity. One should also wonder what is possible based on the actual research situation one is in, but also based on conditions to be imposed on the indices that will be used. Most agreement indices that have been developed are omnibus indices. The index does not make distinctions among various types and sources of disagreement.

In most agreement studies, a classification by two or more raters is only available for a limited sample of units. It is repeatedly desirable to be able to generalize the conclusions to a larger set of units, either those for which only one classification is available or those that will be classified in the future. Statistical methods allow doing this provided that the units on which agreement is calculated are a random sample from this larger population of units. The argument is that when the amount of agreement among the raters is that high that the reliability criterion is met, it does not matter who did the coding, the raters are interchangeable. Therefore, it is allowed to divide the remaining units among the raters. This can save a lot of costs.

Generalizations to other raters are also feasible, provided that the actual raters for which one has data are a sample from a larger pool of raters. In many other applications, however, the raters are fixed and not chosen at random. See the section on raters later in this chapter.

In case the same units have been classified more than twice, however, some more care is required in defining the type of agreement one is looking for. Several views on agreement are possible. This is based on the line of thought that judgments by a majority of raters are expected to be more reliable than judgments by a

R. Popping, *Introduction to Interrater Agreement for Nominal Data*,
https://doi.org/10.1007/978-3-030-11671-2_3

Table 3.1 Assignments by three raters for four units

	Rater 1	Rater 2	Rater 3
Unit 1	pol	pol	pol
Unit 2	ec	ec	ec
Unit 3	pol	ec	ec
Unit 4	pol	ec	none

pol Politics; *ec* Economics; *none* Not coded

single rater. The views are denoted as *mean pairwise*, *simultaneous*, and *majority* agreement. Let us consider one unit. In the first situation, the amount of agreement is based on the proportion of pairs of ratings of the unit into the same category. Table 3.1 contains four units; these are four sentences from articles in newspapers. The three raters coded whether a sentence is about politics, economics, or neither of the two. With respect to units 1 and 2, we see that all three pairs of raters (Rater 1–Rater 2, Rater 1–Rater 3, and Rater 2–Rater 3) have assigned the same category to the sentence—the first sentence is each time coded as being about politics and the second as about economics. As in all pairs, the assignments are identical, the contribution to agreement in 1. With respect to the third unit, we see that one pair of raters (Rater 1–Rater 2) has assigned the category politics, and that the other two pairs (Rater 1–Rater 3 and Rater 2–Rater 3) coded a different pair of categories. One of the three assignments is to the same category, so the contribution to agreement is 0.33. If in each pair the unit is coded into a different category, as holds for unit 4, the contribution to agreement is zero. One speaks about simultaneous or strict agreement in case *all* raters have coded the unit into the same category. This situation is found with respect to units 1 and 2. Hubert (1977: 296) denoted this type of agreement as De Moivre's definition of agreement.

Majority agreement contains a weaker view, and this type of agreement holds if and only if at least k of the m codings per unit is into the same category (here the number *k* is chosen by the investigator). In this example, only at least two out of three raters are possible, so majority agreement is found in units 1 thru 3. We consider only pairwise agreement. An explanation follows later in this chapter in the section on different views on agreement. At this moment, it should be clear that different views are possible. The situation in which units are rated twice is a special case of all three views—here all three work out in the same way.

For this reason, the focus is first on the situation in which each unit is rated twice. This might be twice by the same raters. This allows to conclude on differences in assignments by these coders after agreement has been computed. It is also possible that each coding is performed by different raters. This can happen, for example, in a research situation where the occurrence of the target behavior takes place at very irregular times, or simply where the coding task is so huge that already for practical reasons several raters are needed.

Another important issue in the computation of interrater agreement has to do with the time point at which the category system is established. Usually, this is done before the classification task begins: The raters are given a set of categories and have to assign the units to these categories without freedom to change categories. This type of coding process is known as the *a posteriori method of coding* (Montgomery and Crittenden 1977) (coding comes after the categories are known).

In some cases, however, the task of specifying the most appropriate categories is left to each separate rater, who him or herself has to develop a set of categories during the process of classification. Finding a meaningful grouping of motives into categories may then depend on the kind and frequency of motives found in the answers. This type of coding process is known as the *a priori method of coding* (Montgomery and Crittenden 1977) (coding begins before the category system is completed). In this a priori method, it is important to have at least a subset of units coded by two or more raters and to express in an index whether they agree. Now, however, there can not only be differences in the assignments, but raters may also have developed partly or wholly different category systems. This situation is often encountered in pilot studies where the main goal is the development of an appropriate final set of categories to be used in the real investigation. It is obvious that here it should be known which rater has assigned a category to a unit. Comparing the assignments by the raters of each unit will not contribute to the determination of agreement, but focusing on the assignment of pairs of units does: Has the pair been assigned to the same category as developed by each of the raters or not? The chapter on indices contains a separate section for this research situation.

Agreement is always computed with respect to *one* variable. Computing average agreement across a number of variables is not realistic: Each variable measures a different characteristic. Sometimes, however, variables are used in which it is asked to present more than one answer. An example of such a question is: 'Which are your three best qualities?' The investigator might afterward look at whether quality A is mentioned or not, quality B is mentioned or not, and so on. These are now simple dichotomous variables. The investigator might also want to look at combinations of qualities that are mentioned and at whether these are recognized by the raters. This is the situation of multicoding. For this situation, a measure is available.

The chapter continues with a discussion of the main entries that can affect the amount of agreement that will ultimately be found. Units, categories, and raters have a crucial role in the coding process. These entries need a lot of attention in a study's design phase. An interrater agreement study must be carefully designed. The design involves the number of raters and units that are to be selected and how this is done. The raters constitute a subset of the rater population available for the study, but they might also be the first set of raters that is trained for the task. The raters must be qualified to perform their task. In particular, when raters are supposed to be interchangeable, these requirements are highly important. The same will hold for the units that are sampled to participate in the study. They need to be representative for the population of units. To make it more complex, the entries are not isolated, but usually also are intermixed. This becomes especially visible during the coding task. If the entries are not considered in a proper way, this might not only affect agreement, but also later analyzes in a study. After these entries have been discussed, we can look at requirements to be posed to agreement indices. In the section of requirements, it will turn out that a correction for chance agreement is important. Therefore, the following parts will discuss the different views that exist with regard to the definition of chance agreement. At the end of this discussion, the reader will know how agreement is to be computed; therefore, next the step from agreement to interrater reliability can be made. Agreement so far is treated as agreement between two raters. There are many research situations, however, in which the design is more complex. The most occurring situations are treated next. A more theoretical extension is also possible. This has to do with views on agreement as was mentioned before—only one of these views is recommended. At the end of the chapter, it is shown how agreement is computed when the data are at an ordinal or interval level of measurement. We have to start with the units, the categories, and the raters.

3.1 Units, Categories, Raters

Measurement error denotes the discrepancy between the 'true' score of a unit and the score assigned to that unit. Such errors might have substantial implications on the results of an investigation. This is relevant for the validity of the study. The occurrence of such errors must be prevented. Most potential problems can already be taken care of before the coding task starts. But some errors only become visible while coding is performed. These can be errors that apply to any study, but it might also be that they only occur in a specific type of study.

In his description of reliability Holsti (1969: 135) had mentioned three aspects. These aspects can be dealt with before the coding process starts, so the rater should be prepared:

- Raters' skill, insight, and experience;
- Clarity of categories and coding rules which guide their use;
- The degree of ambiguity in the data.

Disagreements become visible in the off-diagonal cells of an agreement table. All off-diagonal cells in Table 1.1 that contain at least one assignment indicate that raters made a different decision. Krippendorff (2008) distinguishes two types of disagreement: systematic and random disagreement.

Systematic disagreements exhibit some regularity and are more or less predictable. When raters are asked to code candidates' appearance on television in a political campaign, they might disagree based on their own preferences. This can be explained, however. Unless coding tasks are fairly mechanical, rater idiosyncrasies and preferences invariably enter individual judgments and bias the data. These disagreements tend to encourage errors of commission, and an occurrence of the target behavior is recorded that actually is not there. Random disagreements are the ones that exhibit no regularity whatsoever and therefore cannot be predicted or explained. They cannot be considered biased, but fogged or blurred. They are based on incidental misclassifications by raters. These disagreements tend to encourage type II errors, increasing the likelihood of omissions and failing to recognize the target behavior.

The coding process is most relevant in a study because here the data are created. In studies, in which the computation of agreement becomes relevant, these data are usually not delivered by respondents, but are created by people who work for the investigator. This makes the investigator should have means to realize that the data will be of the highest possible quality. There are three entries in a study the investigator can really control. These are the units (that what is coded), the categories (the qualities that can be assigned to that what is to be coded), and the raters (those who have to do this all). Control is already possible before the coding task starts, but also during the coding task problems might come up. The investigator has to be aware of this fact. The entries were also already mentioned above by Holsti. In case agreement is not perfect, it is due to one of these entries or to a combination of entries. The three will be discussed hereafter. All foreseen difficulties or problems should have been considered before the actual coding starts.

3.1.1 Units

In the coding process, two questions regarding the units are relevant. The first one is whether the units are recognized correctly and the second is whether they are understood properly. The answers are to be presented by the raters. Before coding can start, however, there is another important task: the definition of the units. This can be different per type of study. Let us look at this part first.

If one is willing to compute interrater agreement or any other statistical measure, several assumptions should be satisfied. One of these assumptions is that the units are independent. Generally, the units to be coded are easily recognized. In many studies, it is just a person. The difficulty is in recognizing the attribute that is to be coded. An attribute might be found once with a person. Therefore, the question becomes: Is the attribute available or not? If the answer is yes, it might be that one has to ask how the attribute shows itself.

Units as sampled and units as coded might not be the same. The two types are discussed in the context of text analysis and of observational studies. In text analysis, most investigators are looking for co-occurrences of themes in texts. A co-occurrence is usually interpreted as a relation between themes. In case one theme occurs at the beginning of the text and another one at the end of that text, there is a co-occurrence, but it is doubtful whether there is a relation other than co-occurring in the same text. To overcome this problem, texts in a text analysis study are usually broken down into smaller units. This smaller unit is a single group of words regarded as complete in itself. Often this is a paragraph, a sentence, or even a clause (subject + verb + object). Parts that constitute the basis for analysis are called units of analysis. The unit of analysis is that which one wants to characterize. The actual choice of the unit is primarily determined by the theoretical and practical purposes of the investigation. Often the main heading (including sub-heading) of an article on the first page of a newspaper is taken as unit of analysis. One's theoretical argument is then generally that this heading contains all information that was considered by the newspaper's editors to have been most important during that day. Of course, a practical argument for the choice is that the heading is short and easy to identify.

In text analysis studies, three relevant types of units are distinguished: sampling units, recording units, and context units. Sampling units are units that are distinguished for inclusion in the analysis. They are the parts of a text population that are assigned unique numbers and are sampled with known probability. The part of the sampling unit that is really used is the unit of analysis. These units that are also known as text blocks have physically identifiable boundaries. In other fields, similar labels might be used. In observational studies, a similar distinction is possible. The behaviors that are observed are the recording units, but the person who shows these behaviors is the sampling unit.

Recording units, also called coding units or text units, are 'the specific segment of content that is characterized by placing it in a given category' (Holsti 1969: 116). In encoding text, raters record whether or not text segments have one or more attributes. Such text segments are recording units that, for example, might be indicative of aggressive tendencies, economic subject matter, and so on. Recording units can be repeated measures within a sampling unit.

Context units indicate the body of text to be considered in characterizing a recording unit. In manual text analysis, the context unit is usually not made explicit. Here, contextual information is sought when deemed intuitively necessary.

In agreement studies, the attention is concentrated on the recording units. Hereafter, the notion 'unit' in the context of agreement in a text analysis study refers to the recording unit. In the analysis part of a study, it might be important to remember that several recording units can be nested in one sampling unit and that this must be considered during the various computations.

In observational studies, one is usually not only interested in the fact whether a specific behavior is exposed, but also in the moment at which it is exposed and in whether over time there are patterns of behavior. For this reason, the time the observation lasts is split in intervals and the investigator looks at whether the behavior of interest (by each of the participating units [persons]) takes place in an interval or not. It might also be that one only looks at the interval in which the behavior is first exposed. Here are many choices. The behaviors are recorded per time interval; therefore, they can be considered as recording units. In the agreement study, these intervals are the ones that are examined. Sometimes it occurs that the investigator wants to know whether that what is observed (usually a person) shows the behavior during the observation, and the moment at which this occurs is not relevant. So here that what is observed (actually the sampling unit) is the unit.

In medicine and psychiatry, the sampling unit usually is the patient. Here behavior the patient shows is recorded, like the ability to stand up from a chair alone, or the presence of an attribute, for example, a disease. In many cases, sampling and recording unit fall together. When doing observations, this often is not true. Here, a person (the sampling unit) can demonstrate several behaviors (recording unit). A possible difficulty here concerns the definition of the recording units. This occurs in text analysis and observational studies.

Useful recording units must be known. After that, the real coding can start. In text analysis studies, this might mean that first a main characteristic must be recognized (and a recording unit), and next the specific type of this characteristic must be coded. It is possible that also other characteristics of the recording unit are to be coded. The data in Table 1.1 before are from a study where people motivate themselves and others by persuading them that their actions are possible, impossible, inevitable, or contingent. One is to be persuaded with respect to some rationale. In the study, the initial set of categories for classifying rationale types consisted of 'political, economic, cultural, and security-related.' The reality claims have no meaning without a rationale. Therefore, the encoding of data for the reality claims required interrater agreement on the level of the clause at each of the following four steps:

- Identification of a reality claim;
- Identification of the reality claim's rationale;
- Classification of the reality claim as possible, impossible, inevitable, or contingent;
- Classification of the rationale according to a fixed coding scheme.

The clause containing the claim is found by looking for specific verbs, for example 'can' to express possibility. However, one also has to consider the context. It should be a person who has the possibility to do a certain something. A machine can also perform some activity, but this is not a choice, the machine always performs the activity. Recognizing a rationale is different. It has to be recognized from the complete sentence.

Table 1.1 contains the data that remained after the four decisions had been performed. The study started with the first three and the last three paragraphs from 183 texts (sampling units), having 4787 sentences. Only 487 of these sentences were relevant for the study (recording units). Therefore, the computation of agreement is based on the coding of these sentences. In case coding would concern all sentences, agreement would be very high because already about 90% of the sentences would according to the raters fall in the category *not coded*. (It would also show the distribution of sentences over categories is really skewed. This has some undesired consequences as will be shown later.) In particular, recognizing the first part of the coding task, the identification of a reality claim, the raters did very well. They only disagreed in 19 + 12 = 31 out of the 4787 situations.

Recognizing a unit might be difficult in some studies. With respect to a text analysis study, one can express that the unit of analysis is a text or a sentence. These are clearly recognizable. Within such a unit, more than one issue might occur that needs to be coded. In an observational study, this might be more even difficult. Where to start and where to finish an observation? Is an observation determined by fixed time intervals? Should it take place within the interval, or should it only start within the interval. Or does it start when some behavior of interest starts and does it stop when that behavior ends? In observational and medical studies, one often uses videotapes. Now, different behaviors can be scored separately or only one person can be followed. In case the tape contains an interview, the rater can repeat parts to make sure that sayings are understood correctly. This method is an indirect measurement. Not only what has occurred can be caught in a proper way, all raters also receive the same material to judge.

In numerous projects, the reliability study is also used to test that the raters are really interchangeable. The raters really must have operated independently and agreement is to be higher than a specific agreement criterion that has been specified. In case the results are good, it does not matter which rater does which coding, so each rater codes part of the remaining texts.

For the test, actually two subsamples are needed: one for the training and one for the 'examination' part. Units for these samples might be selected at random, but some investigators prefer to include the widest possible variety of units and so to have the selection of a set of units that typify the full range on the variables under investigation. This wide variety of units might be very good in the training phase, but when the final agreement tests are performed one should use a random sample from the complete data set.

In studies where reliability is used to determine whether raters are interchangeable, the investigator often has a subsample of the complete data set coded by all raters. With respect to the size of this subsample, only rules of thumb are available. Neuendorf (2002: 158) mentions a range from 10% to about 20% of the full sample. Lacy and Riffe (1996) want to base the requested number on a desired level of sampling error. An index based on too small a sample size is very unstable. As an example, say 20 units are each judged twice. Each unit that received identical judgments contributes 5% to the amount of observed agreement. But, when the judgments are different and two categories are used, it means that one category is used 5% more by one rater compared to the other and the other category is used 5% less. Therefore, in the situation of two categories, the sample size should as a 'rule of thumb' minimally be great enough to ensure at least 10 observation units in each diagonal cell, assuming the categories are equally preferred (Noda et al. 2006). The minimal number of units does not guarantee adequate power, for which an even larger sample size is required. Methods to compute necessary sample sizes, in the context of inference, have been proposed by Flack et al. (1988) and by Cantor (1996).

Missing data should be possible. It might be that not all units have been rated the same number of times. This might be due to a lacking rater, but it might also be a consequence of the design of the study. Sometimes an observation can only be made if a specific behavior is shown and the moment at which this happens is unpredictable. When the behavior takes place, it can only be scored by the people at present at that moment. As far as these people are raters, they can assign a category to the behavior. An example of a study in which this might happen is the study of symptoms of an attack that is shown by patients in a psychiatric hospital. If such an attack does not happen too often, the investigator will be happy when it happened at night and nevertheless two raters were available.

Relevant characteristics of units that become only visible during the coding process are the following:

- The unit is in motion or not. If so, then coding becomes more complicated. A text is not in motion, but a child that is playing is. This trait is found especially in observational studies and is discussed in some more detail hereafter;
- The characteristic under study is visible or not. If it is not visible, the target behavior should be looked for in another way: being told or explained by the unit (e.g., patient in hospital tells where the pain is) or interpretation by rater (read between the lines in a text);
- Units are large; for example, in a text analysis study, the transcripts of a whole TV program should receive one code (sampling unit = recording unit) instead of a code for one sentence (sampling unit contains several recording units). The same might occur in case in an observational study the time intervals are long;
- Only a small part of the unit is relevant for the coding. In observation studies, one might investigate the occurrence of one specific attribute; other activities are not relevant. In a text analysis study, only the heading of an article in a newspaper is used instead of the complete text of the article.

With respect to observation studies, some more points can be added that are typical for the situation in such studies. By observations, I mean systematic structured observations, not the free observations as generally performed by an anthropologist in the field. Structured observation is only possible when the investigator has specific activities and interactions in mind that have to be observed. He also has to be aware of the types of events which can occur. In such a study, a set of observation categories is used. A famous example of such a set in the field of social psychology is the one by Bales for coding the interactions in small groups (Bales 1950). Bales uses as the unit of observation the smallest unit of behavior that has an independent meaning and that can be perceived and interpreted by the rater. This is often denoted as an act. Bales studied the interactions in a laboratory situation. It is also possible, however, to conduct the observations in a natural setting, where the raters collect the data while the participants engage in their 'normal' behavior.

In observation studies also interval registration is applied. The observational period is divided into short periods of time and the rater should mark whether a given, strictly defined behavior does or does not occur in such an interval. In this context, 'occur' can mean: start, finish, or continue in the interval. The length of an interval is often set to 15 s. As a general rule, it holds that the smaller the length of the interval, the more accurate the encoding. Here, the interval is the unit of analysis.

In literature on observations, four sources of bias are distinguished with respect to the unit (Hollenbeck 1978: 84):

1. Conspicuousness of the observer. Data collected in an unobtrusive way, such as with a video recorder, are different from the behavior collected by a rater present in the situation;
2. Individual differences present in the units themselves. In case the unit is a person, one might think of personality variables like gender or age that might influence the coding by the raters;
3. Personnel attributes of the rater which may determine changes in the unit's behavior. This only occurs when the rater codes on the spot and is visible to the units (persons) being observed;
4. Rationale for observation, reasons for observing units and the way these reasons are communicated to the subject may influence behavior.

At least three additional types of problems can be added to these sources mentioned by Hollenbeck and to the problems mentioned earlier. The problems concern (1) the medium that is used to make observations available; (2) the use of tools to record the target behavior: that what is observed usually moves. This makes recognizing more complex; and (3) in many situations, several different target behaviors are distinguished. Several of these problems can already be overcome in the design of the study.

One has to consider the way data are collected: in vivo or from video (direct or afterward) or by using a mirror. It is often believed that when coding in vivo is applied the rater actually sees more than when the observation is shown on a video screen, but units might anticipate on raters and change their behavior. Video has the advantage that the observation can be repeated many times.

Kent et al. (1979) report on simultaneous observational recordings that were made in vivo, via an observation mirror, and via closed-circuit television. For each form of coding, three raters were used who had extensive experience recording behavior in that way. Raters recorded nine categories of disruptive behavior for children, like touching, playing, noise, aggression, in a special class setting. Frequencies of behavior recorded in these ways differed significantly for only one category: vocalization. This occurrence was found far more frequently in the in vivo situation than in the other two situations. There were no significant main effects or interactions involving the raters' previous experience. Reliability coefficients computed within and between media demonstrated the similarity of observer agreement in all three media. This makes it look as if the way in which the data were collected hardly matters.

Due to technical aids, it is not too difficult today to record the exact moments a specific behavior starts and ends. Several potential problems can already be overcome in the design of the study. Delay (pressing the button late) will only happen if the reviewer does not immediately click on the detected behavior because it was difficult to recognize it. This may be related to the complexity of the behavior itself, but also to the visibility of the behavior (e.g., if a test person is with the back to the rater). In case beginning and end of the target behavior are known, it is easy to indicate whether this behavior occurred in a specific time interval. The investigator, however, has to indicate how the occurrence of the target behavior should be measured. Does one speak about an occurrence as soon as this behavior is demonstrated, no matter how much time it lasts? Or should it last for at least some time because otherwise it might have been unintended behavior, which is not the purpose in the registration? What is the minimum time the behavior should have lasted within a time interval? If two different raters push the button to indicate the occurrence of the same target behavior, but the moment they pushed is slightly different, the investigator can decide it is the same observation and can indicate which time is to be recorded. Investigators might have coders start with a warm-up period in order to make that at the start of the real observation they are fully prepared.

The second problem has to do with the fact that living persons are observed; these do not sit or stand still. Instead of persons also animals or moving object might be investigated. Here, it might be that the target behavior is difficult to recognize. Imagine it is based on a composite score, for example inappropriate behavior composed of a number of behaviors such as noncompliance, hitting, and stealing (Hartmann 1977: 104). It might also be that the rater cannot see the behavior due to the arrangement in the research area. Recognizing behavior should be part of the training of the raters. Not seeing behavior might be solved by placing raters at different places in the area. If the behavior is recorded by one of the raters (or by a majority of the raters), it should be coded as an occurrence. The question whether all occurrences of the target behaviors are recorded is especially important for the list that is produced containing all sequential behaviors. In case time intervals are used, the missing of the behavior is coded as a non-occurrence. In case intervals are not used but instead the sequence of occurrences or not, then missing one behavior might cause that it looks as if there are many disagreements between the raters.

The third problem is met when several behaviors should be coded. In the situation of coding in vivo have to catch all target behaviors at once. This is very difficult even in case the behaviors are easy to recognize. Raters might confuse behaviors or simply push the wrong button. In the situation of pen and paper recording, they might tick the wrong behavior. If they also have to record the time, another source of error comes up. In case observations are coded from video, it is possible to repeat the observation and to code each time another type of behavior. In this way, a lot of mistakes can be prevented, but more time is needed (and therefore there will be higher costs). When the separate codings are combined, it can turn out that several different behaviors all occurred within one and the same interval.

Possible disadvantages of video recording are that often the rater does not see the same things as when coding in vivo is going on. Often the video recording is done by using one camera, so it is always possible that specific behavior will not be recognized. Cameras in each corner of the area might help solving this problem.

Today computer programs are available to record all time aspects of behavior: the beginning and the end of the target behavior (this actually includes frequency), distribution of this behavior over time, occurring simultaneously, sequence, and so on. Several differences might be found: length of a time interval, correction in an interval (is the correct behavior coded—and has it been recorded), interrupted registration, complexity of coding categories (see before, besides often all types of behavior under investigation are sometimes coded at once), and observer drift. With respect to the coding of the behavior that is examined, five possibilities are distinguished:

1. The target behavior is assigned correctly;
2. Previous to the target behavior the rater can wrongly assume behavior. An extra behavior is assigned;
3. The rater does not notice the target behavior and skips it;
4. The rater assigns a wrong category to the target behavior;
5. The rater has determined to assign a category to the target behavior, but in the process of assigning the category is assigned to behavior than his target.

In many studies, the coding is based on observations presented on video. The rater has to push a button connected to a computer and the computer program records the moment at which the behavior under study starts. In this way, finding the sequence of behaviors is in principle not that difficult. One has to consider reaction time by the rater. The biggest problem is the missing of an occurrence.

In case the time is not known, one might have to look at the sequences of occurrences of behaviors. Now identical patterns should be found for the raters. Problems will especially rise when a rater missed a specific behavior. In case at a given moment two different behaviors are coded, the patterns remain identical except for this one behavior. All what happened before and what happened afterward is coded in the same way. This is easily captured. But in case behavior is not noticed by one of the raters, it will look as if all following codings are different. Now, one needs a method that will point to the fact that one rater probably missed an occurrence of behavior (of course it is also possible that according to the rater no target behavior occurred at that place).

The three types of problems mentioned before are not isolated, they also might occur in combination with problems listed in the previous paragraph. An investigator has to take the problems into account. Some can be avoided by acting before the coding starts. All problems affect the coding process, and due to this affect, the amount of agreement that will be computed after coding has finished.

There have been some first attempts using methods of pattern reconstruction to make that the sequence of events will be the same for all raters. Dijkstra and Taris (1995) proposed a method to assess distances between sequences of behaviors. It is based on the number of moves needed to turn one sequence into another sequence. This number of moves is standardized by relating it to the number of moves needed to turn the sequence into its reverse. Weighting it in different manners by the number of common elements in both sequences yields three different agreement measures, suited for different kinds of research questions.

Simon (2006) has another solution to the problem. She assumes that the act that one rater has missed can be recognized. This act should be coded in a separate category: ‘not coded.’ The number of categories is increased by one, and the diagonal cell for this category is empty. Therefore, it does not contribute to the observed agreement, but as at least one of the marginal frequencies is above zero, it contributes to the chance expected agreement. By doing so, the final estimate of the coefficient will not be overestimated.

Also, in text analysis studies, problems like the ones mentioned here might occur. Table 1.1 in the introductory chapter contains a category 'not coded.' This means that according to one of the raters the behavior of interest, a reality claim, is not in a specific sentence, while according to the other rater the claim is there and therefore the type of claim is relevant. Investigators have to decide on how to deal with this problem.

In case not only the occurrence of the target behavior plays a role, but also the duration of that behavior, more issues need attention. For example, does moment in time refer to a specific time block where each block covers a fixed amount of time or does it refer to the moment at which a new action is started? Is the investigator interested in the number of times the behavior is demonstrated within a time interval or is the fact whether it is demonstrated or not already sufficient? Should the behavior concern a reaction on behavior by some other person or not? Other potential problems that have an impact on agreement include time sampling error, situational differences, and temporal instability (e.g., Kazdin 1977). In the assessment of data quality, it is important to identify these sources of error. To facilitate discussion, however, we will concentrate only on raters as the source of error. This is with the full recognition that behavioral observation is a multidimensional situation and that discussion of rater agreement, reliability, and so forth presents only one part of the picture.

3.1.2 Categories

In discussing categories, it is again necessary to distinguish between possible problems that can be considered beforehand and problems that turn out during the coding process. Before, categories were indicated as an exhaustive number of exclusive groups of units that are in some way similar. This implies that in each category a reference is made to a specific characteristic the unit possesses. The categories used in Table 1.1 denote several types of reality claims. These are distinct types without any ordering; therefore, they can be placed on a nominal scale. The complete set of types used constitutes the possibilities for the variable 'reality claims.' In case the requirement of mutual exclusiveness of categories is not fulfilled, the rater might have to choose on grounds that are not correct. In case a set of categories is not exhaustive, the rater might have to code a unit into a category that resembles most, but that is not correct. This problem is often easily solved by adding a category labeled 'other.'

The variable that is at the basis of a set of categories can be very complex. A text analysis study might contain a variable that measures populism in texts. Do we then talk about topics that are characteristic of populists? These can be listed. Or are we talking about the two groups that are most important to populists: we, the own group versus them, the anti-elite? First, these two groups must be recognized by the raters as actors, and next, it should be decided upon whether the language is used in the context of populism or not. It is up to the investigator to give the raters instructions on when which decision applies. These instructions cannot be exhaustive, the researcher can give headlines, and the actual interpretation is up to the rater. Here already is a source for disagreements in coding.

Coding instructions, as just mentioned, are not only relevant for the raters themselves. They are also relevant for other investigators. The instructions enable to reproduce the data making effort or add to existing data. Reliability tests then establish the reproducibility of the coding instructions at different places and different times. Here, different raters are employed, who work under diverse conditions, none of which should influence the data that these coding instructions are intended to generate.

A set of categories might be classified along two dimensions (Hollenbeck 1978: 85): molar versus molecular and extensive versus intensive. This is especially found in observation studies. Molar behaviors are relatively large global units of behavior. Smaller, minuter units are classified as molecular. Molar category definitions are sources of unreliability because they may be subject to greater ambiguity and varying interpretations. Molecular categories may be so narrowly defined that raters have difficulty determining which of the several categories should be used to code a given behavior.

Intensive/extensive describes variation in the degree of detail found in category systems. This is also labeled as complexity. Here, complexity refers to the relative number of coding categories and the distribution of behavior over these categories during a standard observational period.

Recognizing categories usually is related to this complexity of categories. Here follow three examples of a code for behavior used for observing children in a classroom (Romanczyk et al. 1973: 177):

- Out of chair—Observable movement of the child from his chair. When not permitted or requested by teacher. None of the child's weight has to be supported by the chair;
- Modified out of chair—Movement of child from his chair, with some aspect of the body still touching the chair;
- Touching other's property—Child comes into contact with another's property without permission to do so.

Another example pointing to this difficulty is presented by Mikhaylov and colleagues who study party manifestos. These investigators asked raters who participated in a large project on party manifestos as well as a selection of staff and postgraduates in European and North American universities to code data from the training test. Raters coded the units of analysis and all quasi-sentences in the text. The investigators discarded the bottom fourth of the raters in terms of their reliability while dropping none from the top. The obtained agreement coefficients—using Scott's index for multiple raters, to be discussed later—ranged from 0.31

(24 raters, $N = 144$) to 0.40 (12 raters, $N = 72$), which is very low. The results showed that the raters systematically miscoded quasi-sentences, often by coding 'left' issues as 'right' or reverse. No systematic differences were found among raters with different levels of experience. Substantial differences, however, were found in agreement among different coding categories. Raters generally agree on assigning sentences to some categories (such as 703: Farmers), while others (such as 303: Government and Administrative Efficiency) are plagued by misclassification (Mikhaylov et al. 2012: 85). This example shows that the problem of disagreement is not only with the raters, but also with the complex nature of the coding categories.

A similar problem in which few categories are used for coding is described in Popping (2015). In a poll, the respondents were asked which party they voted for in the last elections. This was followed by an open question, asking for a motivation of the choice for this party. The answers were coded into one of the five ideological types as proposed by Converse (1964): Ideologues, Near Ideologues, Group Interest, Nature of the Times, and No Issue Content. Although the categories refer to something one can only imagine, Converse did describe them well. The raters were well trained. Here, the information on the categories provided by Converse received a lot of attention. Also using Scott's index, but now for 2 raters, the amount of agreement turned out to be 0.86 ($N = 2173$). This shows that good results are possible even when the categories are complex.

Complexity of the category system might also be caused by the fact that the number of categories to choose from is large. In the party manifestos project, this number is 56. This becomes especially problematic in case several different characteristics are to be coded at once. '… [C]omplexity can be due to the large number of categories to choose from, but also to the number of different behaviors that are scored within a particular observational system on a given occasion' (Kazdin 1977: 144).

Complexity is also found when categories are made more specific. The molar code 'walk' is easier to deal with than the molecular codes 'step,' 'walk,' and 'run.' Also, a code like 'speak' is easier than 'ask a personal question.' Therefore, differentiating between such codes needs to receive a lot of attention in the training of the raters. But the investigator also needs to be aware of the problem that might occur; often, it becomes only visible during the coding task. Usually, this is too late.

In case the nominal scale is used, the categories are independent, exclusive, and exhaustive. In case the set of categories is not exhaustive, the rater might have to code a unit into a category that resembles most, but that is not correct. In case the requirement of mutual exclusiveness is not fulfilled, the rater might have to choose on grounds that are not correct.

With respect to observation studies, a distinction is made between an ideographic behavior paradigm and a nomothetic-trait paradigm (Suen 1988). According to the first paradigm, only directly observable behaviors can be measured. Researchers using the second paradigm attempt to develop observational measures of psychological constructs which cannot be directly measured. In my view, this distinction does not have to be restricted to observational studies. Instead of behaviors, attributes might be looked at.

Deciding on a unit under the ideographic paradigm can be very difficult even when it is directly noticed. Take the following sentence, which might contain a reality claim:

> What did he get from the mailman? A check that could be cashed in a bank in Graz, for which he *was able to* buy a gambling automat or playing machine that was written off there (in Austria) already (Népszabadság, July 17, 1995, p. 11).

Do we have a reality claim (possibility) or not? The answer becomes clear from the context. Is the author describing the increase in someone's purchasing capacity by virtue of having received a check in the mail, or is this a reference to the person's intention to buy something that only became possible once the check had arrived? The latter would suggest its being a modal usage; the former would not. The coding might also be difficult in case the categories contain vague concepts, such as 'sustainability' or 'globalization.' These attributes are all qualities of the unit that is to be coded. There is no discussion between the raters on the content of the category.

Bennett et al. (1954) argued that the difficulty of achieving agreement increases with the number of categories available for coding, and in the absence of knowing the population proportions of various categories of coded data, the logical probability of categories is the only justifiable baseline to correct the percent of observed agreement.

Coding tasks are ordered on complexity. Crittenden and Hill (1971: 1078), coming from the field of text analysis, distinguish three types: A, B1, and B2. These types have been summarized: 'Type A coding tasks require a coder to find a specific answer to an explicit question at a given place on an instrument. Type B1 coding tasks involve locating relevant information within a larger context …, type B2 coding tasks are those where the coder has not only to locate relevant information, but also to evaluate the relative importance of two or more possible responses to arrive at a single code' (Montgomery and Crittenden 1977: 236). Crittenden and Hill were looking at human coders, who also had to develop a category system. The types can also be used when the categories are already available, and the manifest or latent content of texts is to be coded by humans or by a computer. The tasks are becoming more complex. This will have consequences on the decisions made during the coding task. As this task becomes more complex, agreement possibly will decrease. This has been investigated for text analysis studies: 'The more complex, wide range and inclusive a category is, the more likely that manual coding will find more instances of applicability for this category. However, for categories that are easy to operationalize using a limited number of specific words and expressions, computer content analysis rarely fails in finding these cases' (Linderman 2001: 107).

The nomothetic-trait paradigm usually asks for an interpretation by the rater. The score on intelligence might be used to point to possible different interpretations by raters. Intelligence usually consists of three components: a verbal, a mathematical, and a motor skill. Raters, however, may differ in their view on the contribution of each part. For one rater, the ratio might be 40% for each of the first two and 20% for the last one. For another rater, the contribution of each might be equal, i.e., 33%. This makes intelligence a social construct that is a weighted composite of several attributes. The construct is an artifact in a society which exists because people have accepted it as if it exists and that is measured in different ways. Due to this, disagreement in assignments might be expected.

In many situations, a category is based on a social construct. Such constructs are abstract, intangible, and unobservable and therefore need an interpretation. They are the operationalization of specific concepts, which have meaning in the context of a theory that they are a part of. The DSM handbook (Diagnostic and Statistical Manual of Mental Disorders), the standard work for any professional who makes psychological or psychiatric diagnoses, tells how to determine whether the quality that is investigated is at hand or not. However, in the rules to be followed, the term 'often' is regularly found. This term is not defined any further; therefore, it is not allowed to exclude different interpretations of that quality. In particular, when the nomothetic-trait paradigm is followed, it should be clear that the units as measured are units as seen by the rater.

A characteristic of categories, often found in medical and psychological research, is that one category might be coded more often than the other. This is due to truly unequal frequencies of occurrence of categories within the population under study or a tendency among coders to identify one category more often than the other. The characteristic might also be referred to as a lack of variability. A situation like this usually is undesired as a consequence can be that now chance expected agreement is as high as observed agreement. Up to here, the agreement index was implicitly defined as the proportion of agreements on the diagonal of an agreement table. Better is to say this proportion divided by the theoretically maximum proportion of agreements, which is one. Some agreement is to be expected due to chance alone. In Sect. 3.2, this is discussed in detail. Correction for chance would imply that this chance expected agreement is subtracted from both the observed agreement in the nominator and the maximum possible agreement in the denominator. In case chance expected agreement is as high as observed agreement, the final index will take the value zero, as this is the value found in the nominator. However, there are studies in which the detection of a relatively rare event is the primary objective. A situation like this is often found in studies on characteristics that scarcely occur. An example is found in the already mentioned study by Spanjer et al. (2009). It did not happen too often that patients in the investigation had activity limitations. The situation where one category is used far more often than the other categories is known as prevalence.

This situation might also be found in observation studies or in studies dealing with the diagnosis of rare diseases. In observation studies, one might work with fixed time intervals and a specific behavior might only be found in a small number of intervals, just because it does not occur that often.

Say two raters who had to code whether something applies or not agree completely in their assignments, see the two situations in Table 3.2. In the situation where the quality *does not apply,* both raters code 98% of the units in the one category and the remaining 2% in the other category. The amount of observed

Table 3.2 Tables with skewed distributions

A		Rater 2			B		Rater 2		
Applies		*Yes*	*No*	#			*Yes*	*No*	#
Rater 1	*Yes*	2	0	2	Rater 1	*Yes*	0	1	1
	No	0	98	98		*No*	1	98	99
	#	2	98	100		#	1	99	100

agreement is 1; in anticipation on the presentation later, the amount of chance expected agreement based on the two marginal distributions (in this situation: the distribution of the units over the available categories as found for each of the raters after coding is finished) is 0.2 × 0.2 + 0.98 × 0.98 = 0.96. But as the proportion of observed agreement is one, it follows that the final index also takes the value one. Now assume that with respect to these 2% in the other category, one rater coded one unit in the first category and the other in the second, and for the other unit, they just did reverse, as is shown in situation B in Table 3.2. Now, one would find that the observed agreement is 0.98, and the chance expected agreement is 0.01 × 0.01 + 0.99 × 0.99 = 0.98. The final index will get the value zero as the proportion of observed agreement equals the proportion of chance expected agreement. The index that is computed tends to underestimate the agreement on the rare category. This despite the fact that the amount of observed agreement is also very high. This situation is indicated as a paradox (Cicchetti and Feinstein 1990).

For this problem, a solution is necessary. Later, we will see that this can be looked for in several directions. One, sometimes heard as a statement, is that if one category occurs only few times, it is a rare case, and if it occurs it turns out to be rated in different ways, then this fact must be punished severely, so it is very good the index takes a low value. This is especially relevant in case the investigator wants to say something about these rare cases.

Another problem, connected to the labeling of the categories, can be that raters had difficulty in distinguishing between different categories given these are at a nominal level of measurement. In case the categories are ordered, it might be difficult to decide where one category ends and the next one starts. Here, one might look at intercategory or interclass agreement: How often does it occur that two specific different categories are assigned to a unit?

So far it has been assumed that the categories to be used during the coding process are known to the raters an advance. The raters applied the a posteriori method of coding; this is the far most common research situation. It is possible, however, that the raters also have to develop a set of categories by themselves. This implies that differences in coding might not only be due to differences in assignment to categories identical for the raters, but also to different views on how the categories should look like. Such situations sometimes occur, mostly in the pilot phase of a study, where the final categories are still under discussion. Here, the a priori method of coding is used.

Having such data, one can also compute agreement. The indices, however, are not based on the assignments per unit, but on the fact whether the raters assign a pair of units to the same category (as developed by each rater), or not. Four situations are possible: (1) both raters assigned the pair to the same category; (2) the first rater assigned the pair to the same category, but the second rater did not; (3) the first rater assigned the pair to different categories, but the second one assigned the pair to the same category; (4) both raters assigned the pair to different categories. This last possibility implies that the raters agree in the view that the two units in the pair belong in different categories as developed by each of them. The two units really indicate different issues. This situation is further discussed in the part on the indices.

The number of categories available also plays a role with respect to the amount of agreement that might be found. The fewer categories are available, the more the raters have to confine themselves to this limited number of options. Therefore, it will be 'easier' to assign the same category to two different units, which will result in a higher value for the amount of observed agreement. In case chance expected agreement is also to be considered, the impact of the number of categories will also become visible in another way, this is when this number is considered as the basis for the distribution of units over categories. This view will be discussed when chance expected agreement is an order. Most important is that one should use the number of categories that are needed given the research question and design. One should be aware, however, that the more the categories, the lesser the frequency in (some of) the categories.

In case the set of categories is large, say 20 and there are two raters, then there are 20 ways to agree and 380 ways to disagree. Here, the investigator might worry whether there are enough units to give the raters sufficient chance to disagree relative to the possibilities.

Darroch and McCloud (1986) claim that in many circumstances the (ordered) categories into which units of observation are classified do not have precise objective definitions. Consequently, one has to recognize that different raters may interpret the category definitions differently and that categories will not be completely distinguishable from each other, even by the same rater. These investigators used ordered data. For such ordered data, the claim might be true, more interesting is whether it is true for data at a nominal level of measurement.

Later, we will see that it is possible to compute agreement for each separate category. This can be useful in finding the category that causes most problems to the raters. The investigator has the possibility to verify whether some unforeseen or unintended problem is going on.

3.1.3 Raters

Raters should be well trained for their task. Requirements they should possess and the training program are discussed later. Being well trained does not mean there is uncertainty with regard to assignments that are made. However, it is actually always assumed that when raters assign a code they are completely certain about their choice. We cannot imagine that a rater makes a choice at random. However, there will be many situations where a rater takes the position that alternative *a* might be the correct choice, but this might also hold for alternative *b*. Alternative *c* for sure is not the correct choice. The rater will follow some reasoning based on which *a* or *b* is chosen, but the rater knows this choice might be wrong. There are hardly studies in which this problem is addressed. A consequence, however, is that part of the agreement found is essential, i.e., not due to chance, and a part is not essential (so, due to chance). Some authors (Aickin 1990; Gwet 2008) have developed measures taking this fact into account. They needed for each assignment additional information, namely is the rater convinced or is a random guess made. Normally, this information is not available. Therefore, I will not address this point any further. Other investigators claim that explicit models of rater decision making are needed to solve the problem (Uebersax 1987). Here, the investigator would need both a theoretically defensible model and sufficient data to empirically verify conformity of observed data with the model. Anyway, *modeling* rater agreement will be needed (Agresti 1992; Uebersax 1993), so far this hardly took or takes place.

Apart of this, three topics are relevant when raters are considered. The topics are the rater's skills, the fact whether the raters are known or not and the view on how they use the categories. The first topic concerns the rater's skills. The most common research situations in which coding is performed are the ones in which raters are used who are in principle equally skilled. These are 'common' raters performing a text analysis or doing observations on specific behavior that is shown mainly by all kinds of people. The raters had a training in which they learned how their task is to be performed. Later aspects of such training will be discussed. After the training, the raters are supposed to be well qualified, equally trained, and therefore equally skilled. This makes that their coding is reproducible and that basically the raters are interchangeable. The distribution of categories over all raters is the only stochastic variable there is. Now, two positions are possible. One position is that the actual raters that will perform the coding are a sample from a larger pool of raters. The other position is that the raters are fixed and not chosen at random. This has consequences for the way in which chance expected agreement will be computed.

An alternative view, especially important when the nomothetic-trait paradigm applies, is that each rater has specific knowledge or qualifications. This knowledge or these qualifications are not available with other persons; therefore, these raters are not interchangeable, and the distribution of units over the categories is dependent on the rater. This view we find often in psychiatric research. Here, the context of association is at present, and there are as many distributions as raters.

There are a few situations, however, in which the raters possess very own specific qualifications, which makes them unique to perform the task. Such raters cannot be interchanged. A psychiatrist is an example of such a type of rater. The person has qualities in recognizing phases in highly complex states of the mind of patients and progress in development in such stages that can only be performed by a few people. Each psychiatrist has an own way to conclude regarding a patient, and this method cannot be generalized.

In many investigations, a rater performs the complete coding task. This is absolutely necessary in case a rater has own unique qualifications that might play a role in the classification process. When this condition of unique qualifications does not hold, which actually is the case in most research situations, the exact rater is not relevant as long as the raters are equally skilled. More relevant is that the units are by preference rated the same number of times and anyway more than once. There are even settings in which one rater cannot perform all codings. Say one studies symptoms shown by patients in hospital having an attack of epilepsy. As it is not predictable when such an attack comes, the coding must be performed by the persons who are there at the moment it occurs. Each time, these will be different people.

The way how often categories are used as well as the fact whether the rater's own skills are relevant has consequences for the way chance expected agreement is computed. This agreement is based on the frequency in which each category is used. In the literature, three views dominate the discussion. One of these views is elaborated in two different ways. The presentation of these views starts in the next section.

Before it was argued that in case the number of categories used is small, observed agreement might be high just because the rater has not that many possibilities for assigning. In this situation, however, also the likelihood for raters to agree by pure chance increases very much. This is because both raters must confine themselves to the limited number of options available, which impacts the overall agreement rate, and not necessarily their propensity for essential agreement, i.e., agreement not due to chance. Therefore, the joint probability of agreement will remain high even in the absence of any essential agreement among raters. A useful interrater agreement coefficient is expected to be close to zero, when there is no essential agreement, and to increase as the essential agreement rate improves. Most chance-corrected agreement coefficients achieve the first objective. However, the second objective is questionable, simply because it is not known how certain the rater was when a code was assigned to the unit.

One might say that the more often a unit is coded, the better. In the ideal situation, all raters operate in an identical way, so there will be no differences in assigning units. In case still some miscoding occurs, this error will be overruled by the other correct assignments. This is not realistic. Higher agreement can be seen as a stronger identification for reliability. Practice is that the more the raters, the more the disagreement. Each rater makes sometimes mistakes and each does this at another place; therefore, more raters will contribute to more mistakes. Now, it is a good sign when the amount of agreement found is nevertheless above the criterion for reliability. This is an important issue. Another problem is that coding takes a lot of time and is expensive. Besides trained raters should be available. For these reasons, one will often be happy to have at least two raters. I always try to find three raters, but in most studies this turned out not to be realistic.

Before, it was already indicated that some authors emphasize the relevance of standard raters. In some situations, it may be reasonable to assume a 'correct set' of codings, to which an investigator might wish to compare the codings of other raters.

The one who produced this set of codings is considered as the standard, sometimes even the 'gold-standard.' For example, when a medical doctor or psychologist believes that he has correctly classified a sample of patients into a number of symptom categories, he may wish to compare the categorizations of his students with his 'correct' assignments. The standard rater is an expert in the field. Landis and Koch (1975: 157) suggest that '"true values" may be created by using a separate panel of experts to provide a standard classification for each subject.' As the standard's assignments are the basis for comparison, it is assumed that the standard does not make misclassifications. It is up to the investigator to decide whether this assumption is correct.

In case sample size small, the computation of expected agreement can be biased because a minor difference in marginal can have enormous consequences. The only solution I am aware of is presenting only the amount of observed agreement and come with a warning.

There are some other problems in which raters are involved that only become visible during the coding process:

- Systematic errors occur when there is a rogue rater. Although rarely encountered, there is always the possibility of the appearance of a rater who simply cannot—or will not—be trained to achieve reliability;
- Features of the rater that have an effect on the units, such as gender or age in life observations;
- A rater will work more accurately when it is known that his or her scores will be compared to the ones by another rater (Romanczyk et al. 1973);
- Rater drift. There is a shift in the criteria used by a rater: Raters '"drift" from the original definitions of behavior. Drift refers to the tendency of observers to change the manner in which they apply the definitions of behavior over time' (Kazdin 1977: 143). An example is a change in coding behavior due to a change in definition, it is possible that aggression is first coded to occur after five hits and later on already after four hits. This should be covered in training or repeated training;
- Features of the rater that have impact on the coding process, like being tired or lack of motivation;
- Expectations raters in observation studies have regarding sequences of behavior, it can be that a specific behavior is usually followed by another specific behavior, but this does not always have to be the case;
- Feedback raters receive from the investigator in relation to the expectations regarding the unit's behavior (Kazdin 1977: 147).

It is the investigator's task to prevent that these problems do not occur. Knowing the relevance of units, categories, and raters, it is time now to look at the agreement indices themselves. First one has to look at requirements to be posed to such measures.

3.2 Requirements to Be Fulfilled by Agreement Indices

Empirical investigators have not paid much attention to the statistical properties of agreement indices. According to Bartko and Carpenter 'too often a reported reliability reflects an association between pairs of raters rather than an accurate assessment of agreement between ratings. Furthermore, practices such as: (a) devising special reliability measures for a particular study which precludes scientific generalizability and therefore does not permit comparisons with other studies; (b) failure to take chance expected agreement into account; and (c) reporting reliability without an associated test of statistical significance are commonplace' (Bartko and Carpenter 1976: 307). The last point asks for a comment. When agreement indices are at issue, inference, especially the test of significance, is generally used to test whether the estimate that is computed deviates from the value zero, a value one gets when all data are assigned at random. With a large number of units, it is practically always possible to find that the two values differ. Attention for the confidence interval might be more interesting now. This comes back later in the text.

In an agreement study, only units can play a role that have at least been rated twice. If only one coding is known, no agreement with respect to the unit involved can be determined. The total number of units included is denoted by N. Using γ to indicate a type of index in general, the simplest index is:

$$\gamma_0 = O,$$

where O is the observed amount of agreement between the raters. Raters always operate independently. The minimum value this index can take is 0; the maximum value is equal to the number of units. Indices of this type can hardly be compared; therefore, they are normed so that their values are in a certain range. Several types of indices can be distinguished. The first one is:

$$\gamma_1 = O/M,$$

where M is the maximum possible amount of observed agreement, this usually is found when $M = N$, the raters agree on all assignments. The index takes values in the range 0 to +1. Rogot and Goldberg (1966) denote this proportion as the 'index of crude agreement.' The index is an important descriptive statistic and has a unique commonsense value. Earlier in the text, this index was used with respect to the data in Table 1.1: $\gamma_1 = P_o/P_m = 0.88$. In case only complex statistics are reported, the reader might not be informed at a practical level. Hollenbeck (1978: 87) list three advantages and five weaknesses of the index. The advantages are: (1) ease of computation, (2) apparently simple interpretation, and (3) the sensitivity to systematic errors. The weaknesses are: (1) the magnitude of the proportion scores is dependent upon the marginal probabilities in the agreement table between two observers or an observer and a standard; (2) the proportion scores provide no information about sources of measurement error; (3) the proportion agreement computed upon total trial scores reveals nothing about agreement on component parts; (4) proportion agreement varies as a function of the time interval used to define time blocks; and (5) as the proportion has no metric properties, it is not amendable to further mathematical analysis. These points that hold for observation studies have to come back some way hereafter. But more is to be said, another disadvantage of the index is that in cases with few categories in the nominal scale the probability of equal assignments is greater than in cases with many categories. This is especially important when prevalence, the popularity of a specific category, plays a role. All disadvantages in some way reflect to the fact that the magnitude of percentage agreement can be inflated by chance. Many indices of this type γ_1 are known as similarity indices as used in cluster analysis. In case the index takes the value 0, this does not indicate the absence of agreement but is evidence of maximum disagreement. No agreement at all is statistically rare and occurs when raters follow incompatible coding instructions.

Some investigators do not start from the observations on the diagonal of the agreement table but start from the marginals. The summed proportion over the categories of absolute differences between corresponding marginals indicates the amount of agreement (Woodward and Franzen 1948). The outcome zero is the perfect outcome. Flanders (1967) therefore uses 1 minus this amount. The problem is that equivalence of corresponding methods does not mean equal assignment of the same unit.

A second type of index is:

$$\gamma_2 = (O - 1/2M)/(1/2M),$$

where the possible outcomes are in the range −1 to +1. Here, a correction for the range of the index is made. In case the index shows a positive outcome, there is at least agreement on half of the units that have been coded. In the final type of index, a correction is made for an agreement that is expected on the basis of chance:

$$\gamma_3 = (O - E)/(M - E).$$

where E is the amount of agreement to be expected by chance or under the null model of no agreement. In general, there will be already some agreement among the raters even when they assign their categories at random. Different views exist on how to define this null model; the discussion on this issue follows later in the text. Assume two categories *applies* and *does not apply* that occur the same number of times. In case two independently operating raters would assign completely at random 100 units to either category *applies* or *does not apply*, one might expect that 25 units will be assigned by both raters to category *applies* and also 25 units by both to category *does not apply*. This implies that the amount of agreement is already (25 + 25)/100 = 0.5 only based on random assignment. A correction of the index is necessary to prevent the impact of this unwanted amount of agreement.

A beneficial characteristic of the γ_3-type index is that it allows for generalizability across different experimental conditions. Foster and Cone (1986) pointed out that chance agreement changes as the base rate of behavior changes. Because percentage of agreement does not correct for chance, it is differentially inflated in situations with different rates of behavior, hindering comparison across conditions. γ_3-type indices, however, allow for standardized comparisons by removing chance.

Usually, the amount of chance expected agreement is based on the proportion in which each category has been used by each rater, across all raters or by the standard rater (in case such a rater is available). This will be discussed later. This index takes the value 1 in the situation of perfect agreement and the value 0 if the observed agreement is equal to the amount of chance expected agreement.

The index takes a value less than 0 in case the observed agreement is less than chance expected agreement. The lower limit is $-E/(M - E)$; if $E = ½$ M this lower value is −1. This last type of index is to be preferred, as will be explained later. Cohen (1960) defined his index of this type as 'the proportion of chance expected disagreements which do not occur, or alternatively, it is the proportion of agreement after chance expected agreement is removed from consideration.' Note that in case $E = ½ M$ the indices γ_2 and γ_3 are identical. Often the constituting parts are also written in terms of proportions. In that case, the index is usually written as: $\gamma_3 = (P_o - P_e)/(1 - P_e)$, where P_o is the proportion of observed agreement and P_e is the proportion of chance expected agreement.

Not all investigators believe that correcting for chance is useful: 'Agreement is an absolute measure and … RELiability is a RELative measure to quantify observer variation. Cohen introduced κ as a measure of agreement and many authors have followed suit as the need to take chance agreement into account sounds quite plausible. Adjustment of the observed agreement for the expected agreement, however, turns κ into a relative measure. Therefore, we should stop adapting Cohen's κ and instead be critical about whether a specific clinical question asks for a measure of reliability or a measure of agreement' (De Vet et al. 2013: 3). The index κ is a type γ_3 measure that will be discussed later. The authors responsible for this remark look at agreement as a descriptive entity. Reliability does not matter. The focus is on the judgments in the situation where agreement among the raters was found. The presence of random agreement that is always there is ignored. Even if assignments are compared just to see whether the raters differed, some reliability criterion like 'should we be satisfied with this result or not' still plays its role. At a given moment, one might not be satisfied with the outcome, this is only possible when these results are too low. Implicitly, there is a criterion value for the agreement. Let this value be explicit and be based on a correction for what always will be there, so for chance.

Sometimes the chance-corrected index is presented in terms of disagreement. Say Q_o is the proportion of observed disagreement (all units that are not on the diagonal in a table like Table 1.1) and Q_e is the proportion of chance expected disagreement. As all is presented in terms of proportions, we know $Q_o = 1 - P_o$ and $Q_e = 1 - P_e$. Now, it is easily seen that $\gamma_3 = 1 - Q_o/Q_e$.

Later, we will see that four different views on chance exist. The most restrictive one assumes uniform marginal probability; all categories have equal probability to be used. A less restrictive assumption is marginal homogeneity; the distribution of units over categories is the same across raters. Uniform distributions for each of the raters are not required; there are only identical distributions across raters. Here, chance can be based on using replacement or not. The estimate of the occurrence of a specific category, given independent assignments by the raters, is equal to the mean occurrence of that category: the sum of the number of times it was used by each rater divided by the number of raters. The least restrictive assumption includes the requirement of independent assignments by the raters but no marginal homogeneity or marginal uniformity.

The requirements mostly found in the literature are in general terms or for the measurement of comparable notions. These will be presented here for agreement indices for nominal data, disregarding whether they are obtained by means of the a priori or the a posteriori method of coding. Use is made of the requirements for statistical indices that are found in the literature. This literature will not be considered in detail.

Investigators have to convince their audience that these assumptions are satisfied. Many methodological problems in testing reliability stem from violating these assumptions. First, the requirements will be mentioned. Next, explanations will be given with regard to some of the requirements; more details are given in Popping (1988). In most studies, the situation is that the raters are known and that they are interchangeable. This makes that the exact identity actually is not relevant. Whether this fact applies or not has enormous consequences. This becomes already visible in the formulation of the second requirement below. The issue is elaborated in the section of recognizing raters. The requirements are:

1. The maximum possible value of the index is 1, regardless of the number of raters or categories;
2. In the case of independence given the frequency in which each category is used by each rater or over all raters, the index takes the value 0;
3. Each rater agrees perfectly with himself or herself;
4. Perfect agreement is a transitive relationship;
5. Permutations of categories may not lead to other results. Because the data are measured on a nominal scale, the order of the categories is arbitrary and does not influence the results;
6. The estimated value of the index is independent of the number of units;
7. If there are more than two categories, it should be possible to compute the amount of agreement for all categories together, but also per single category;
8. If there are more than two known raters, it should be possible to compute the amount of agreement for all raters together, but also per single rater;
9. The index should be symmetrical with respect to the raters or ratings involved. An exception can be made for the situation in which one of the raters is taken as a standard;
10. The sampling distribution of the index or at least the variance should be known, at least approximately;
11. The index should be robust;

12. The index should be simple and interpretable;
13. The index should be valid.

The first two requirements are the most important ones. They need some explanation. First, the maximum possible value of the index is 1, regardless of the number of raters or categories. In his study on models of statistical relationship, Weisberg (1974: 1639) distinguishes three types of maximum relationship. Only the first type is relevant in agreement studies: 'strong monotonicity,' each category of one variable is uniquely connected to a category of the other variable. With respect to agreement, this should also be true. The other two types are 'moderate monotonicity' and 'weak monotonicity.' The first one is about adjustment for an unequal number of variable categories. The other one states that the categories of one variable are as homogeneously as possible distributed over the categories of the other variable.

The marginal distribution is the distribution of the frequency in which each separate category is used by a rater. In the case of perfect agreement, the marginal distributions of each of the raters are equal, so there is no doubt about the maximum. But how to define it if there is no perfect agreement? The maximum can be defined as dependent or independent on the marginal distribution by each rater. In the latter case, the maximum is reached when all units in an agreement table are on the diagonal of that table (given the sequence of the categories is the same for all raters). In the former case, however, the reasoning is that a certain cell on the diagonal can never contain more units than the lowest frequency in the corresponding marginal, so this frequency is a determinant for the maximum. This view was presented among others by Brennan and Prediger (1981), Umesh et al. (1989), and Von Eye (2006). It comes closest to weak monotonicity.

As an illustration take the data in Table 3.3. We have 100 units that are coded by two raters into one of two categories: Some quality applies *yes* or *no*. Here, it concerns a quality that rarely applies or that rarely does not apply. The first rater coded 80 units into category *yes* and 10 into category *no*. For the second rater, these numbers are 80 resp. 20. The observed agreement P_o among the raters is (80 + 10)/100 = 0.9. Using the minimum of the corresponding marginal totals, one would find for the maximum possible agreement P_m = [min (90, 80) + min (10, 20)]/100 = (80 + 10)/100 = 0.9. The ratio P/P_m is 0.9/0.9 = 1. This ratio will not change after a correction for chance expected agreement is made. The corrected amount in nominator and denominator of the γ_3 formula is always the same, no matter how it is computed. In this way, it will look as if the agreement is perfect, which is not correct. The disagreement on 10% of the assignments is not

Table 3.3 Table with skewed distributions

		Rater 2		
	Applies	*Yes*	*No*	#
Rater 1	*Yes*	80	10	90
	No	0	10	10
	#	80	20	100

considered. The maximum is the maximum that is possible given the data as they are available, but this maximum is far from what is possible.

In the case of agreement, perfection is demanded: The categories assigned by the raters must have an equal content; there is no question of predicting a category from one of the raters from a category given by another. This means that only the maximum that is independent of any marginal distribution is of importance. In terms of proportions, this maximum is 1.

The second requirement states that in the case of independence given the marginal distributions, the index takes the value zero, which means there is no agreement other than expected by chance. Indices can take extreme values, either because there is no correction for the actual minimum value they can take or because there is no correction for the results that can be expected based only on independence (Mokken 1971: 55).

The concept of 'no relationship' is discussed by Weisberg (1974: 1640–1641). He presents four definitions and elaborates these for dichotomous data. Here, the elaboration is presented in terms of the agreement and is for multiple categories.

1. 'Accord,' the outcome zero is found because at least one rater has assigned (nearly) all units to one specific category;
2. 'Balance,' the outcome zero is found because the proportion of units on which agreement exists is equal to the proportion for which this agreement does not exist;
3. 'Independence,' the outcome zero is found because the proportion of units on which agreement is found equals the proportion that is to be expected based on chance;
4. 'Cleavage,' the outcome zero is found because the distribution of units according to one of the raters is equally distributed over the categories as used by the other rater. This is shown in Table 3.4.

In the situation of a 'cleavage' relation, the requirement of an 'independence' relation is also fulfilled, and assuming a 2×2 table this is also true for the 'balance' relation. One should not want the 'accord' relation because there one rater has assigned all units to only one of the categories. A backdrop of the 'balance' relation is that the number of categories has an impact on the proportion of units on which there is no agreement. The 'independence' relation is in line with the definition of independence that is used in this study. The 'cleavage' situation is a special case.

Table 3.4 Example of a 'cleavage' relation

		Rater 2			
	Category	*1*	*2*	*3*	#
Rater 1	*1*	10	10	10	30
	2	20	20	20	60
	3	30	30	30	90
	#	60	60	60	180

The index γ_3 contains a correction for chance expected agreement. In the choice of a transformation formula, the demands 'null in the case of independence,' 'fixed maximum,' 'fixed minimum' are in general conflicting; the last demand is usually dropped. Arguments for this are that the minima often concern a pathological situation, while agreement less than expected by chance is not interesting for criteria concerning the content, and hardly ever occurs. Very often extreme results are possible (Mokken 1971: 56). These extreme results are removed in indices containing a correction for chance. Therefore, indices of this type are preferred. A disadvantage of these indices is, however, that they cannot handle some very extreme situations: In the unusual and in general undesirable situation that a rater has assigned all units to one and the same category, it is not possible to investigate the difference between observed and chance values. These will be identical as there is nothing to choose from. One can wonder whether in future assignments there will also be at least one rater who codes all units into just one category, 'the measurement is not sensitive enough to warrant any decision about constancy' (Galtung 1979: 111). In practice, this problem will hardly ever occur. The best is to norm for chance, given independence. Later on, it will be explained this is not that simple, as several views on independence are possible.

Often the raters are equally skilled persons and it is not relevant which raters performed the coding task, the ratings are interchangeable, but sometimes raters have very specific skills that others do not have and that are relevant for the task. There should be no hesitations to enter the ratings for each unit into the database in random order. If there is any such hesitation, then it may well be that what is seen may not be unequivocally interpretable as agreement or disagreement. In such situations, the context is association, not agreement (Bloch and Kraemer 1989: 272). This occurs, for example, if one of the raters is considered as a standard or if one rater is female and the other is male and it is assumed that gender has some effect. Whether the context is agreement or association has its impact on the way chance expected agreement is considered. The context, however, might also be reflected immediately in the requirements mentioned before: If one rater is a standard, the index need not be symmetrical. This comes back when raters are discussed.

The requirement that each rater agrees perfectly with himself or herself is based a discussion about E-correlation coefficients by Vegelius (1978: 606) and perfect agreement is a transitive relation comes from Vegelius and Janson (1982: 244). Quade (1974: 370) takes the position that a certain class of transformations *f* (e.g., the class of linear functions f or the class of monotone increasing functions *f*) of the data cannot lead to other results: $c(x', y') = c(x, y)$, with $x' = f(x)$ and $y' = f(y)$. Vegelius and Janson (1982: 243) and Galtung (1967: 209) also mention this. Galtung remarks that for data at the nominal level all permutations and all distinct recodings are permitted. As the data are at the nominal level, the sequence of categories is arbitrary and does not affect the results.

Galtung (1967: 209) claims that it should be possible to compare the outcomes found by using an index with other outcomes where the same index is used. The same position is taken by Janson and Vegelius (1982: 244): An index must not be dependent on the number of units N. This requirement is fulfilled when the results are presented for proportions.

Some authors claim the index cannot be used for making comparisons between outcomes in different studies. This is because there are 'gradations in terms of symptomatology or the salience of diagnostically relevant information' (Uebersax 1987: 141). This is demonstrated in examples in which the probability of true scores is used. In most agreement studies, certainly within the social sciences, the true scores are not known.

Mokken (1971: 55) mentioned that it should be possible to investigate the scalability for all items together, but also for each separate item. For agreement indices, this would mean that it should be possible to compute the agreement per category but (under certain conditions) also per rater. Both types of indices are only useful in case there are more than two categories or more than two raters. Mokken (1971: 60) also mentions a number of conditions that concern the values of coefficients in the population. Galtung (1967: 209) also highlights this point: The sampling distribution of the index must be known. This is discussed in the section on sampling characteristics. However, we will not be detailed with regard to population values and sample characteristics of the indices.

In the literature, especially in Galtung (1967: 208–211), still a number of distinct criteria are distinguished which are important, but which in practice are very difficult to be assessed. The first one states that a parameter should be stable. It is not allowed that the parameter is sensitive for small changes such as moving a unit from one cell to another cell in the agreement table.

The next desideratum states that it should be possible to calculate a measure in a way as simple as possible and that it should have statistical and substantive significance. The measures mentioned before consist of some components: observed, maximum, and chance expected agreement. This makes it easy to understand how the index is constructed. However, the definition of especially observed and of chance expected agreement can be very complex based on the actual research design. The fact that chance can be expressed in several ways contributes even more to this. But it should be possible to get a clear picture of the components.

The estimate of the index that is computed is not the truth. Therefore, inference should also be a requirement. If the variance of the estimate is known, it is possible to perform a significance test, but also lower and upper bound of the estimate can be computed.

With respect to statistical significance, one could consider a correction for a maximum, minimum, or expectation under a zero relationship. Though, this does not have a social scientific meaning. According to Galtung, the ideal would be to characterize a distribution having a measurement as parameter that comes as parameter from a mathematical model. However, this hardly happens. Usually, the interpretation of a given measure is based on a statistical–mathematical basis, without a particular substantive interpretation being present. Mokken mentions a number of conditions coming from values coefficients take in a population. However, he refers as the greatest problem to the question of when a coefficient is sufficiently high for a population (Mokken 1971: 60).

Finally, it was mentioned that an index should be valid. Validity cannot be computed, but a number of demands can be mentioned that should be realized in order to make that the assigning process is as optimal as possible. These demands have to do with the training of raters, but also with the formulation of categories and rules to be followed in the coding process, see also the remarks in Sect. 2.2. The characteristics of a number of elements which play a role in the field of research are measured by using the value an index takes. These elements have distributions. The possible values of the index are determined by these distributions. The structure of these values should be investigated on their measurement qualities (Galtung 1967: 210).

Although it is not formulated as a requirement, we will see later in the text that possible problems in the interpretation of the results of a computation can be avoided if categories are used with more or less the same frequency, at least the distribution of categories should not be too skewed. Note, there is no requirement stating that the agreement index should be independent on the number of categories. On the one hand, observed agreement will probably be higher when there are less categories. But depending on the distribution of units over categories, it is possible that chance expected agreement becomes higher in the situation of less categories and therefore that the value the index takes becomes lower.

A limitation of the γ_3-type index is that in certain situations depending on the way chance expected agreement is computed it is impossible to compute an estimate for the index. This only happens when all units are encoded into only one of the available categories.

Popping (1988) compared 39 indices. Most of these did not fulfill all the requirements mentioned or could not be applied in all research situations. Although the indices are listed at the end of the book, including where they fail, they will not be discussed in the text. A problem for which no standard solution is found is the definition of chance expected agreement. This is related to the role the raters play. It is discussed in the next section.

What is needed to perform a coding task, so to get the data one needs in order to use agreement indices? Crittenden (1971: 229) presents the following list:

1. A set of data;
2. A specification of the size of the units of data to be taken into consideration for each part of the coding task;
3. A set of categories;
4. A set of code designations, one for each category;
5. A set of rules for assigning the data to the categories.

The designations and the rules should guarantee perfects assignments, but practice is that there are always some differences in the classifications by raters. Each unit should be encoded into a specific category. This implies that units should be well defined. The same holds for the categories. Assigning codes is performed by raters, these should be well equipped, but also equally equipped. These three, units, categories, and raters, have been discussed in more detail before. Now, the step to the definition of chance expected agreement can be made.

3.3 Chance Expected Agreement

The amount of observed agreement found usually is too high, because some agreement is already to be expected by chance as we have seen before. Therefore, a correction is necessary for this type of agreement. Chance expected agreement would be the agreement that is found when all assignments are made at random. In that case, the agreement between the assignments by the raters will be zero. Assuming the hypothesis of no agreement among the raters is correct and the samples are random and independently selected, we can compute a set of frequencies that would be expected given these marginal totals. There are restrictions:

- Each random number must be in the range of the number of categories;
- The total number of random numbers equals the number of units times the number of raters, assuming each unit is coded by each of the raters;
- The distribution over categories can further put a restriction on the randomness.

With respect to this last point, four views on distribution over categories are possible: uniform marginal probability, marginal homogeneity with replacement, marginal homogeneity without replacement, and independent assignments. The views are discussed hereafter. After that, it should be possible to decide on which view is to be preferred in which research situation.

The data sets in Table 2.2 show that each total in a row or column in the 3 × 3 table is 30. This makes that the amount of expected agreement in any cell is the marginal proportion by the first rater times the marginal proportion by the second rater times the number of observations is (30/90) × (30/90) × 90 = 10. This would imply that the frequency expected on the diagonal in the table is 10 + 10 + 10 = 30, or as a proportion 1/3. In reality, however, the distribution of the marginals is usually very different from the ones in this example.

3.3.1 Chance Expected Agreement Based on Uniform Marginal Probability

The first view was presented by Bennett et al. (1954), who looked at the answers to open-ended questions in a public opinion poll. The investigators questioned a number of respondents for a second time. It is assumed now that each category has equal chance to be used, and there is marginal uniformity. Therefore, the proportion of chance per category is $1/c$, where c refers to the number of categories that are used. In the situation of two categories, the correction would be 0.5. Now, the index turns into an index of type γ_2. If three categories would have been used, the correction is 0.33 and so on. Chance expected agreement is based on completely random assignments. This proportion 'represents the best estimate of P expected on the basis of chance and discloses the extent to which the empirical score proportion P is affected by the number of response-categories' (Bennett et al. (1954: 221).

The index has reappeared in Janson and Vegelius (1979), Brennan and Prediger (1981), and in the two-category case in Holley and Guilford (1964) and Maxwell (1977). The assumption of equal use of each category, however, is usually unrealistic. Zwick (1988) points to the fact that this index can lead to spurious underestimation of the chance expected agreement rate. This occurs when the number of unused categories is increased. She presents gender as an example. Usually, the categories are 'female' and 'male.' An investigator might add 'hermaphrodite.' The category will hardly or not be used, but decreases the amount of chance expected agreement very much. This will have a positive effect on the value that will be found for the final index. Brennan and Prediger (1981: 689) indicate that chance agreement based on marginal distributions (they restrict themselves to independent assignments, see hereafter) 'clearly depends on the degree of agreement between the assigners' marginal that may result from expert judgement rather than "chance."'

Krippendorff (1987) has used the term 'equity' to refer to this type of correction.

3.3.2 Chance Expected Agreement Based on Marginal Homogeneity with Replacement

The second view was found in the work by Scott (1955). Scott, doing text analysis, needed many raters to analyze the answers to open-ended questions used in survey research. He was just interested in 'check-coding': If one part of the data was coded by one rater, and another part by another rater, would that make a difference? It should not. In research situations, after a serious training, there should be no reason to assume that one rater is better equipped to perform a coding task than another rater. The raters have to code a quality of the units. The population contains one distribution of the units, but it is not known how this distribution looks like. In the long run, it is approached best by averaging across the distributions over the individual raters who code (representative) samples of units: 'It is convenient to assume that the distribution for the entire set of interviews represents the most probable (and hence "true" in the long-run probability sense) distribution for any individual coder' (Scott 1955: 324). This implies that the actual raters are considered as a sample from a larger pool of raters. The raters are assumed to be potentially interchangeable; that is, in the population of units, the m ratings for each unit are intended to have a distribution invariant under all permutations of the indices. This implies the m ratings might be entered in random order. It also implies that all units should be coded a certain number of times, but that this coding does not always have to be performed by the same raters. For this reason, sometimes the term non-fixed raters is used. Scott did not present an agreement table in his text, but he described the task to be performed by the raters by presenting an example. The answers by respondents (units) to the question 'What sorts of problems are your friends and neighbors most concerned about these days?' The available categories referred to the nature of the problem: economic, international, political, local and personal problems, and not ascertained (Scott 1955: 324). In order to be able to perform the coding, the raters should be trained, but do not need very exclusive skills or knowledge. However, each category should have the same denotation for each rater.

In this situation, chance refers to the statistical independence of the set of units—the units that are investigated—and the categories collectively used to describe them. Raters make independent decisions on the same units; therefore, the investigator has one stochastic variable estimated by the mean across raters of the distribution of categories in the data. Scott used the Greek letter π (pi) to indicate his index.

As descriptive statistics, interrater agreement indices describe the degree of agreement between two or more particular raters used in a study. By themselves, they indicate the extent to which those particular raters participating in the assessment are interchangeable. With a high agreement index estimate, an investigator may also wish to conclude that any similarly trained rater (other than the ones used in the reliability check) using the same system will produce similar results. It also implies that the coding task can be reproduced. To make such an inference requires that the raters participating in the actual assessment procedure be randomly drawn from all possible similarly trained raters, usually an impractical requirement (Suen 1988: 349). But they can be drawn from the available raters.

Without knowledge of the correct category to which units are assigned, ‘this conception takes the distribution of values [categories, rp] that all coders collectively use to describe a given set of units as the best estimate of what the population of units is like’ (Krippendorff 2011: 97). Investigators have to use the distribution of categories the units are assigned to. This distribution is a quality of the units. It becomes visible when we consider the distribution over all assignments. It implies that at the end all raters should end up with the same distribution. In case of differences, these differences should be due to random errors and should not be considered.

In case well-trained raters perform the same coding task, for which no special knowledge is needed, the assignments should be identical and we have agreement. As examples take the rationales found in sentences as reported in Popping and Roberts (2009). The sentence:

> Signing a petition is possible, because “today is not too late to stop … government-initiated monopolization of … the media.”

is about politics, and the sentence:

> Our experiences of sickening displays of wealth are possible, because “it is drawn from (black market) resources drawn away from the economy.”

is about economics. The decisions to come to these conclusions might have been difficult and might have asked for specific interpretations, they point to a quality of the sentences. Investigators doing text analysis usually use dictionaries during the coding process. These dictionaries tell which word or phrase as found in a text corresponds to which category that is relevant in the investigation. But the dictionary should also contain rules for handling when decisions are hard to formulate.

In observation studies, one usually looks at whether the target behavior is at hand or not. This might concern activities like: touching, noise, vocalization, orienting, playing, or aggression. In all situations, a quality of the unit is considered. The distribution of these qualities is based on the distributions over all raters. Therefore, the computation or chance expected agreement should follow the line of reasoning as proposed by Scott.

Scott assumes the marginal distributions for the raters are more or less identical. The homogeneity can be tested using a test by Stuart (1955) or improvements of this test. This test is considered later when indices are discussed.

3.3.3 Chance Expected Agreement Based on Marginal Homogeneity Without Replacement

A slightly different view is presented by Krippendorff. Krippendorff (2004) assumes that there is only a limited number of equally qualified raters, these are the raters that participated in the coding process in a specific study. The raters are fixed. Chance expected agreement should be based on the codings by these raters. Therefore, his computation of expected agreement is based on the estimate of the distribution of all codings of categories in the actual data. Therefore, a randomly selected unit should not be replaced after it is drawn from the distribution of available codings of units. This implies that for each draw this distribution will be slightly different. This fact makes that the definition of chance expected agreement according to Krippendorff is slightly different from the definition used by Scott.

Scott follows the principle of statistical independence; there is a complete independence between the chance expected codings. The first choice does not give any information about the second chance. Krippendorff allows choosing only from the ratings that have not been selected yet. It implies Scott does apply replacement of the unit and Krippendorff does not.

Krippendorff masks the fact of independence. He mentions that α and π differ in the factor $N/(N-1)$, and that '… this factor corrects α for the effects of small sample sizes and few coders' (Krippendorff 2004: 419). The difference in outcome between the two indices is an artifact of the way of computing chance expected agreement and ends in the index being not independent of the sample size as was asked in the sixth desideratum. Assume two coders who could use two categories. The observed proportion of agreement P_o is 0.5 and the proportion in which each category was used is also 0.5. For any N, Scott's index will take the value zero. This is different for α. Given four units which are all in a different cell of the agreement table, i.e., the frequency in each cell is 1, α will get the value 0.25. In the situation of 40 units, α becomes 0.025 and in case there are 400 units, the index takes the value 0.0025.

Conclusion: α makes a correction for chance expected agreement, but it is not a correction for chance expected agreement given independence. Due to the way α is computed, the index is not independent of the sample size.

When we sample with replacement, the two sample values are independent. Practically, this means that what we get as the first value, one doesn't affect what we get as the second value. Mathematically, this means that the covariance between the two is zero. In sampling without replacement, the two sample values are not independent. Practically, this means that what we got as the first one value affects what we can get as the second one. Mathematically, this means that the covariance between the two is not zero. That complicates the computations.

Scott assumes that the raters constitute a sample taken from a pool of equally qualified raters who are interchangeable. Krippendorff assumes the raters constitute this pool. The expected agreement is computed from an estimate of the distribution of categories in the population. For the computation of the chance expected agreement, one draws each time a random unit from the distribution of categories over all raters. For each draw, the distribution is identical. This implies that after each draw the unit is replaced.

The indices by Scott and Krippendorff are almost similar, certainly when the sample size is large. Hereafter, only the one by Scott will be referred to.

3.3.4 Chance Expected Agreement Based on Marginal Heterogeneity

In this last view, the previous position was attacked. This was done by Cohen, working in the field of clinical psychology: 'one source of disagreement between a pair of raters is precisely their proclivity to distribute their judgments different over the categories' (Cohen 1960: 41). Here, chance expected agreement is considered as a function of the individual rater's preferences for the categories, not of the estimated proportions in the population of data or of an estimate of the distribution of categories in the data. In this view, the individual preference reflects individual rater bias. That what is noticed by one rater might be different from what the other rater has seen. Cohen only used the above sentence in his article to explain this. In the example he presents, content does not play a role. The categories are indicated by a number. After the presentation of the example, possible categories are listed, however: 'schizophrenic,' 'neurotic,' and 'brain-damaged.' In a later article, Cohen (1968) presents three categories 'personality disorder,' 'neurosis,' and 'psychosis,' but he does not mention that raters might have different views. As we are talking about social constructs here, different views are very well possible. Each category might have a different denotation for each rater. These different views are in line with Suen's nomothetic-trait paradigm, and the unit cannot be measured directly. The relevance of aspects of the disease might be judged differently by one rater than by another rater.

Cohen defines chance agreement as the agreement that would be observed if two coders operate independently, regardless of how often they used the available codes. By not counting raters' unequal affection for categories available for coding as unreliable, this definition fails to treat raters as interchangeable. It numerically rewards them for not agreeing on their use of categories and punishes those that do agree (Zwick 1988). The view is in line with the tradition of analyzing correlations among variables.

Cohen uses agreement to show reproducibility. He starts his first paper by referring to the fact that agreement 'quite parallels in its logic the concept of the coefficient of equivalence used with texts' (Cohen 1960: 39). Each test is different and therefore will have an own distribution of categories. Here, the equivalence relation is referred to: It is possible to predict the scores on one test from the scores on the other test.

The second issue concerns equally (and well) skilled raters. Cohen (1960: 38) formulates: 'In the typical situation, there is no criterion for "correctness" of the judgements, and the judges are a priori deemed equally competent to make judgments. Also, there is no restriction placed on the distribution of judgments over categories for either judge.' Here, the phrase 'equally competent' causes problems, and it allows different assignment procedures. In the examples Cohen presents, different procedures are indeed used. Here, his approach differs from the one by Scott. We have all the requirements that were already listed. Raters are supposed to act in concrete situations in the same way. This should be a part of the training process. In many situations, certainly when professionals are acting as rater this becomes questionable. The raters are competent, but the rules they have to follow often are not defined. In particular, when the categories are social constructs, this might cause problems. The raters can fill in parts in their decision-making process at their own discretion; they possess special skills and use these in their own way. Now, in fact one should ask whether the assignments by one rater can be predicted from those by another rater. This is done when the reasoning proposed by Cohen is followed with respect to chance expected agreement.

An example in line with Cohen's view is the following. Popping (1983) used agreement to determine the extent in which respondents are classified in the same way according to different allocation rules. In this study, each respondent is indicated as mainly having one of four types of work orientation, also social constructs, based on the outcomes of different scale analysis methods. The author investigated on which of four (standardized) scales, each measuring one of the orientations, a respondent ranks highest; next, the corresponding orientation is attributed to the respondent. This all was based on the outcomes of four different types of scaling models. These types can be seen as raters. Each scaling method has its own characteristics and is based on an own mathematical model; therefore, the methods are independent from each other. But each method also uses different (explicit) rules and criteria. Here, the investigator is actually looking for association. This is not association between two variables each having an own set of categories, but between two variables both having the same set of categories. In the example, the variables are two methods. In case the assignments according to the two methods are made visible in a table, one will look at what is on the diagonal of that table. Connected to each method, that follows an own assignment procedure, is an individual distribution of respondents over the categories and this causes that chance expected agreement can be computed based on two distributions. Usually, it is not assignments by methods that are compared, but assignments by raters. The agreement informs on whether assignment by one rater can be predicted from assignments by the other rater, i.e., on whether assignments are reproducible. It does not inform on the question whether raters are interchangeable. The index according to Cohen is known as κ (kappa). The agreement between the final classifications in the example was not high; therefore, one might conclude that the type of scaling method used matters.

In case social constructs are to be coded, the circumstances are similar. Each rater uses an own method or reasoning to decide on which construct is at hand. Psychiatrists use their own (implicit) criteria, although they all follow the DSM. Also, for physicians, a similar reasoning might hold. Deciding on diseases might be based on weighting out all alternatives that are possible in the decision-making process that is already complex because so many characteristics are involved.

The approaches by Scott and Cohen differ very much. This is due to the starting position taken by the two proposers. In a lot of literature, this is ignored. The difference in value the index takes is usually small when it is computed according to both methods on the same data set. The reason for this is that the marginal distributions for each rater turn out to be not that different. In Chap. 4, we will see that the index according to Scott is more conservative than the index according to Cohen.

3.3.5 Chance Expected Agreement Discussion

Based on a Monte Carlo study, Feng (2013) concludes that chance expected agreement according to Bennett et al. is only affected by the number of categories, and that chance expected agreements of π and κ are affected by the marginal distribution, the level of difficulty, and the interaction between them. In particular, the difficulty level has an enormous effect. The more difficult the coding tasks are, the lower are the chance expected agreements. With respect to marginal distributions, two issues are relevant: (1) Where is the distribution based on: the distribution of categories per rater or this distribution over all raters, and (2) how does the distribution look like: skewed or not. This last problem is already discussed. A possible solution can only follow after concrete indices have been considered. This will happen at the end of Chap. 4.

We have seen that basically three different views on chance exist. The most restrictive one assumes uniform marginal probability; all categories have equal probability to be used (Bennett et al.). A less restrictive view assumes marginal homogeneity; the distribution of units over categories is the same across raters. Uniform distributions for each of the raters are not required; there are only identical distributions across raters (Scott and Krippendorff). Drawing units can be performed using replacement or not. The least restrictive view assumes the requirement of independent assignments by the raters but does not assume marginal homogeneity or even marginal uniformity (Cohen).

It was already noted that the first view is not realistic as here chance is based on complete random assignments. In case the marginal distributions are identical, the index according to Cohen gives the same outcome as the index according to Scott. In literature, the view on chance expected agreement as formulated by Cohen has received most attention. More coding tasks have been performed in his field of application than in the field Scott was working in. A consequence is that for most users all indices of type γ_3 as presented above became indicated as κ indices, the name Cohen is used to denote the index. Both Cohen and Scott defined their index for the situation a unit is rated twice. At a given moment, extensions of the index to more than two raters or ratings have been proposed. One method was proposed by Light who followed Cohen's approach. Another method, proposed by Fleiss (1971), claimed to follow Cohen, but actually Scott's approach was followed. It was Conger (1980) who made this difference well defined. He does not explain, however, why the difference is necessary. The same holds for other authors who emphasize this difference (Fleiss 1975; Zwick 1988). Since, however, extensions of the Scott index are sometimes indicated as Fleiss index.

Fleiss has claimed that π seems inappropriate as a measure of agreement (Fleiss 1965: 656). But there is a critique: 'the meaning of "perfect" and "chance" is subject to assumptions that are often accepted *implicitly*. ... Reliability data properly stem from two or more judges who make *independent decisions* on the *same units of observation* and in terms of the *same descriptive categories'* (Krippendorff 1978: 143). The answer is that Krippendorff 'has missed the point of keeping the two judges' base rates ... separate. It is not at all that κ ..., which do[es] so, assume[s] (that) judges draw from *two separate populations*. Rather [the index] assume[s] that judges will inevitably employ somewhat different criteria for distinguishing between the presence and absence of a trait, even when judging members of the same population' (Fleiss 1978: 144). Hsu and Field (2003: 216) prefer κ over π, because 'the assumption of marginal heterogeneity may often be more realistic than the assumption of marginal homogeneity.' It was already mentioned that according to Bloch and Kraemer (1989: 272) such homogeneity is part of association, not of agreement. My reading is that the base rate in the sample is based on a quality of the units that are studied, but in many studies this view is not followed. In these studies, the base rate is determined by criteria posed by the raters. I realize that in some situation, where the coding task is very complex, this can hardly be avoided.

The following is in line with this position taken by Fleiss. Hollenbeck (1978: 91) presents the results of a study in which raters had to perform observations. He claims that in order to compute Scott's π the distribution of coding categories should be known in advance and should be equal for both raters. This is not the situation in the data set he is presenting in his text, and in general, he claims, this will not be true in any observational study. Therefore, computing Scott's index is not correct. He, however, excuses Scott while he is working in the field of text analysis and there these assumptions are more appropriate. Hollenbeck uses practical arguments, not theoretical ones. This is to be regretted. It is also to be regretted that Hollenbeck did not discuss the differences between an observational study and a text analysis study with respect to the distribution over categories of variables after the assignments had been made. Now, he only made a statement, for which he does not deliver any proof.

Actually, in the majority of the situations in which coding is performed, the position taken by Scott is the correct one. That what is to be coded is an attribute of the unit of analysis; therefore, there is one distribution over all units. As this distribution is not known in reality, it is best estimated by using the assignments over all raters. In most research situations, the coding into the correct category can be performed well, although training is needed in order to realize that all raters operate in the same way in the situation of complex assignments as far as these occur. Regrettably, there are hardly data on training.

Coding becomes complex when constructions are used. These are not by nature attributes of the units, but these are qualities developed by investigators or users in the field. In the coding process, these raters, who are in most instances experts in the field, use subjective interpretations. Even, although coding rules have been agreed upon, such subjective parts remain present. Think of decisions on: 'high, medium, low;' 'better, worse;' 'small, large;' 'often, rarely/occasionally.'

Therefore, it is relevant to consider whether the assignment procedure that is applied by each individual rater is identical or different. The discussion on this issue does not take place. Cohen refers to situations in which especially psychologists or psychiatrists perform the coding. These raters usually had a training once, and without doubt, they follow the protocols that apply when assignments are to be made. My guess is that here it stops. The question whether they follow these protocols in an identical way is an open one. A test to find out is not performed, and a training is not at all the issue. Therefore, each rater might decide differently. Raters in text analysis or observation studies occasionally get a training. But details on the quality are not available.

In the previous section, examples of the association situation have been presented. A situation slightly different from the one in which classification based on the outcome of scaling method is performed is found in Light (1971) who compares codings regarding children by fathers and by mothers. Landis and Koch (1977) present a diagnostic classification regarding multiple sclerosis based on assignments by neurologists in two completely different hospitals. It is questioned whether fathers consider other aspects than mothers do and whether in one hospital other rules are followed than in the other one. In both study designs, the association context is at hand, and therefore, the raters are not exchangeable. In particular, in studies where social constructs are used, this might be obvious. Here too, raters are already considered as experts and do not get an additional training.

Before, it was already mentioned that when a rater is compared to some established criterion one speaks about accuracy. Usually, this criterion consists of the assignments by one very qualified rater: the (gold) standard. According to Vanbelle and Albert (2009: 96): '[a] gold standard generally represents some practically not attainable but only approachable level or quantity determined by a single reference method. There are situations, however, where a gold standard may result from the application of several reference methods or the opinions of several experts, without necessarily achieving a perfect consensus on all items.' This view was already met in the field of biology, where Duggan (1992) points out that sometimes it seems that what is investigated implies a level of perfection that in general cannot be attained. Therefore, the concept is approached by using the classifications by several raters as a standard.

The assumption is that the marginal according to the standard is the correct one. Therefore, this marginal is to be used in computing expected agreement and not the marginal by the other raters. These other raters are as they were still in a learning process in order to become just as qualified as the standard. For this reason, the marginal they produce might still contain errors.

We saw that it has been suggested that the scores by the standard might exist of scores by several expert raters. One might ask whether this is good thinking. As Vanbelle and Albert present it, it is an ad hoc solution for a practical problem. No theoretical motives are given. At this place, the main concern is on the way chance expected agreement is considered. The way maximal agreement is found, however, is also problematic.

Vanbelle and Albert base the distribution over categories on the assignments over all raters in the standard, so they follow Scott, i.e., raters are interchangeable. They do not assume, as Cohen does, that these raters use own interpretations in the decision making with regard to each patient in the sample. Next, assignments by an individual rater are compared to the standard. Here, the authors follow Cohen, and they assume that the individual rater will perform the decision-making process in another way than the standard raters do. Why is this so?

Maximum agreement is based on the sum of the lowest frequencies per category as used by each rater in the standard. This implies that here Cohen is followed: Each rater in the standard has an own distribution. This is very confusing; two methods are intermixed.

In many other studies, however, my guess is even in most studies, the investigator is not interested in association. Relevant is whether the raters really end up with identical codings when they all follow the same decision-making process. If the criterion is met, not only the coding process is reproducible, the raters are also exchangeable. These raters need training in order to make that they indeed make the same decisions. I am afraid the relevance of such training is underestimated.

Investigators should be very explicit on their view on the coding process. In part, this is to make that their readers will better understand what did happen, but also to get knowledge with regard to a possible demarcation line between applying the same coding process and applying different processes.

Sometimes the decision on whether there is one distribution for all units or one distribution per rater can be hard. Spanjer et al. (2009), for example, looked at whether patients can perform a certain activity or not. The yes/no coding would fit perfectly in the approach according to Scott. But the yes-answer was split into two or three ordered subclasses on how well the activity can be performed, like 'a little bit' versus 'very well.' At this moment, one becomes confronted with the interpretation by the physician. One example concerns 'lifting or carrying.' Here, the categories were:

0 normal, can carry or lift about 15 kg (toddler);
1 slightly limited, can carry or lift about 10 kg (small toddler);
2 limited, can carry or lift about 5 kg (bag of potatoes);
3 severely limited, can carry or lift about 1 kg (one liter of milk).

These categories are ordered and rather concrete (especially due to the example that is included), so raters should not have too much problems in deciding on the code. Decisions were based on written reports of a semistructured interview with disabled patients on impairments, activity limitations and participation, and on a written report on physical examination. In the report, it should be mentioned how much one can lift. (Note, the amount is mentioned by the patient, it might be that this amount is not in line with reality.) In the study, only the weighted proportion of observed agreement was reported (for weights see at the end of this chapter) because distributions were very skewed and the number of patients was small. Would chance expected agreement have been used, Scott should be followed.

Such discussions on the distribution that is at hand, or better on the assignment procedure, is usually ignored. Investigators just follow one view. There are lots of studies in which Cohen is followed, but where Scott would have been the better choice. The patients reported what they can lift or carry, and their guess, which might be completely wrong, is mentioned in the report. The raters base their assignments on the information in the reports. This information is so concrete that own interpretations are hardly needed.

The view on the number of distributions that can or should be used also has consequences for extensions of the γ_3-index, this can be with regard to the number of raters used, but also on the role raters have. When the Scott view is followed, the change expected agreement is always based on one distribution. This is the distribution over raters, no matter how many raters performed the coding task (although each unit should at least be coded twice).

Looking at what takes place when a coding task is to be performed by humans, one is confronted with the fact that at least two ratings per unit must be employed in order for interrater reliability to be assessed. Just as all variables in a study should be subjected to reliability assessment; all raters whose data are used in the study should be part of the reliability assessment. It is not generally acceptable to use only a subset of raters in reliability assessment.

Raters operate independently, so all coding of the data should be conducted independently, without discussion among the raters. If possible, it is also to be preferred that the raters are 'blind' for the hypotheses and research questions in the study. In this way, this knowledge can have no impact in whatever way on their coding.

Assuming that all raters are equally skilled and are interchangeable, one might say that all raters used the same marginal distribution. If there is disagreement, it is random disagreement. As raters are interchangeable, it is not relevant which rater produced which classification. The data collection may even have proceeded by using different pairs of raters for different units. In that case, one may still use an agreement index for the overall reliability. Then, however, we shall speak of agreement of judgments, rather than of raters.

Investigators should wonder which situation applies in their research and should not blindly follow one of the two approaches. By preference, they should motivate their choice. Now, it is known how agreement is computed. From here, the step to interrater agreement can be made.

3.4 How to Proceed When Agreement Is Satisfactory?

The strength of agreement that is needed for a reliable result in a study is decided upon in advance by the investigator. This strength depends on the importance of the data with respect to (policy) implications based on results found in the study. For simplicity, assume that only the O-part (the proportion of observed agreement) constitutes the index. The value 0.7 now means that on 70% of the units the raters agree in assigning to the same category, while this is *not* true for 30% of the units. It means that 30% of the units is for whatever reason not classified by the raters in an identical way. Note, there is normally no standard telling what the correct way of coding is, the exception being the situation where accuracy is investigated. In the actual research situation, it is to the investigator to decide which value of the agreement index is acceptable. When the correction for chance is made, the index indicates the agreement above what chance would yield. Ordinarily, the value that is found now is less than the proportion of observed agreement; actually, this proportion is biased upward. The chance-corrected index is more severe. Assume for the moment the amount of chance expected agreement is 50%, now the outcome of the index is $\gamma_3 = (0.70 - 0.50)/1.00 - 0.50) = 0.4$. This is far less than the outcome based only on observed agreement. Again, it is to the investigator to decide which level of agreement is still acceptable. Rules of thumb are mentioned at some places. For example, Landis and Koch (1977: 165) take the position that Cohen's kappa should be over 0.80 to be sufficiently good. The authors even present a complete overview, see Table 3.5. According to Banerjee et al. (1999), this is already so when

Table 3.5 Labeling strength of agreement for kappa according to Landis and Koch

Range kappa statistic					Strength of agreement
		κ	<	0.00	Poor
0.00	–	κ	–	0.20	Slight
0.21	–	κ	–	0.40	Fair
0.41	–	κ	–	0.60	Moderate
0.61	–	κ	–	0.80	Substantial
0.81	–	κ	–	1.00	Almost perfect

Source Landis and Koch (1977)

kappa = 0.75. For his coefficient α, Krippendorff (2004: 429) also mentioned the value of 0.80. In general, the decision on reliability is comparable, no matter whether the approach by Cohen, Scott, or Krippendorff is followed.

These rules by Landis and Koch or any other proposer of rules should not be followed strictly. The proposers did not supply evidence to support these choices. The acceptable level of agreement depends on the actual research situation. Some tasks are in some way more difficult than other tasks, and therefore, it might be that one should already be satisfied with a lower amount of agreement. In case one is performing a content analysis study and is looking for manifest content, then in general the coding task is easier than when one is looking for latent content. Here, also the risk of errors of commission comes in, i.e., texts are retrieved that are irrelevant. For this reason, one might even take the position that not the value of the index should be above the criterion in order to conclude that the assignments are reliable, but that the lower confidence interval of the index should be used. Besides, one has to look at the distribution of the assignments over the categories as we have seen before.

A high level of agreement between raters does not guarantee an individually correct classification; disagreement between raters, however, usually indicates a lack of accuracy. This indicates that it is difficult to distinguish between categories or that the training of the raters was not good enough.

A variable that does not reach a researcher's minimum standard for reliability should be dropped from the further analysis. Sometimes it is also possible to reconfigure the variable with fewer and better-defined categories. This should be done during the pilot coding process, prior to the final data coding. When several variables together are needed to support a conclusion, the smallest amount of agreement among these variables has to be used as an indication for the reliability of the whole process leading to the conclusion.

Usually, the conclusion that agreement is high and coding has been performed in a reliable way is not the end of the line, but only a step in between. A high amount of agreement is seen as an indication that raters are interchangeable. Often a subset of the complete sample of units was used for the agreement test. With respect to the units that were not in this subset, it does not matter which rater performs the coding. However, often the investigator uses a design in which there is some overlap between the raters (say 30%), which allows to perform a new agreement test at the end of the coding process. In case the focus of the investigator is on interchangeability instead of reproducibility, the demands for reliability are higher. Here, one might also use another view: The lower bound of the confidence interval should be above the criterion. As samples are used, there is always a small chance on errors; by taking this lower bound, this chance is eliminated.

In case the interrater agreement is not at the appropriate level, the data should be thrown out. Before doing this, however, the investigator has some possibilities to check whether it is really that bad. The first one is to administer an additional data collection. For this, an extra rater is needed, a rater who should have had the appropriate training. But one can also verify whether the actual raters understood the coding procedure or whether there have been circumstances that might have a negative effect on their performing the job. See also the remarks in the section on selection and training of raters.

The determination of the amount of agreement between the assignments by the raters usually is a first step in the data analysis. In case the data turn out to be reliable, the investigator might want to confront the variable under investigation with other variables. The investigator might, for example, want to investigate whether women score different than men, or younger people different than older people. In the situation of two raters, the investigator can continue with the assignments by one of the raters now. The relatively few differences in assignments are supposed not to matter; these differences are randomly distributed over the data set. In case the number of raters is more than two, the investigator might conclude to use for each unit the code that was assigned most. In case two codes were used the same number of times, but more often than other codes, one of the two is taken at random. Given two raters, the investigator might not use the assignments by one of the raters, but in the situation of disagreement between the raters he or she might take one of the codes at random.

3.5 A Number of Relevant Research Situations

So far, the situation was considered in which all units are rated twice; this can be by known raters or not. One index was used that informs on the total agreement among all units. In practice, however, one might be confronted with research situations that are different.

A number of empirical situations are distinguished on grounds that they require different kinds of agreement indices. One important aspect of these situations concerns the coding process. If units are to be assigned to categories that have already been developed, it is said that the a posteriori method of coding is applied. In this situation, all raters use the same set of categories. Another possibility is that the judges have to develop the categories as well as assign units to them; this is called the a priori method of coding. Here, the raters do not necessarily use the same set of categories. For these two situations, different types of agreement indices should be used.

Other aspects of empirical situations that will be distinguished include:

- The identity of the rater per unit is registered (if not the term agreement among ratings is used rather than among raters). This is a requirement in case one wants to follow Cohen;
- The number of raters or ratings. In the situation of more than two raters, it is best to use pairwise agreement, as will be explained in the next sub-paragraph. With respect to the construction of the new variable to use later on, however, one can use the assignments by one of the raters, assuming coding was performed in a reliable way, but one can where possible also use per unit of observation the code used by the majority of the raters;

- All raters (individually or as a group) are compared to a standard rater. This is especially important when the training of the raters takes place;
- One rater is compared to another rater or a group of raters. This tells whether the rater is different from the other raters;
- The best or worst rater is looked for. This is close to the previous point; it tells whether the rater is different from the other raters;
- Two groups of raters are compared, raters in one setting are compared to raters in another setting. Now, the effect of the setting, which might be a training method, becomes possible. For each group, the sample of units must be the same. If not, different samples are compared and therefore other methods have to be used;
- The amount of agreement for a fixed category. This allows finding where difficulties during the coding process are or have been located;
- The amount of (dis-) agreement for two different categories. This also allows finding where difficulties during the coding process are or have been located. The method makes visible whether two categories have been interfered;
- The amount of agreement within a category, computed by taking one of the raters as a standard. Especially useful during training;
- The number of judgments per unit. At least two judgments per unit are necessary. In case Cohen is followed, all units must be rated by the same raters. In case Scott is followed, this is not necessary;
- Missing judgments. For some reason, not all units might be rated the same number of times. If Scott is followed, this is not a problem, but it is if Cohen is followed;
- The number of categories. This number should at least be two;
- Multicoding. A unit can be coded in more than one category.

Deriving computational models for even more situations is not that difficult, the question is whether such situations really are found in reality and are worth considering. In my view, the above-mentioned situations are the ones that occur regularly. Indices that are or will be derived for any combination of these aspects should be very consistent and should be realistic. Developing indices for situations that can in practice not occur, or will not occur, are not needed.

In situations where the raters are exchangeable, it does not matter which rater performs the coding. It can even imply that units are coded by different raters. The only relevant point is that each unit is coded the same number of times. Later, we will even see that being coded at least twice is actually already sufficient. Now, the index as proposed by Scott is to be used; later, we will see that this index can be formulated in such a way that the exact raters do not play any role. Of course, one cannot investigate now that differences in assigning are due to a specific rater. This should not matter, however; the raters are exchangeable due to, for example, the training they have had. An implication is that in case one still wants to have the possibility to check on rater effects, even when these raters are exchangeable it is necessary that all units are coded by all raters.

The use of weights that indicate the seriousness of disagreements will be considered very shortly. As we focus on nominal data, such weights do not play a role. But to serve readers who would like to use weights, they will be discussed shortly with respect to the comparison of assignments by two raters. Besides, there are some situations where weights are used as a kind of trick, for example, as a method to combine as it were categories.

Before, it was stated that agreement can only be computed for one variable at a time. Sometimes this limitation can be overcome by choosing smart categories. Schouten (1993) investigated the presence of geographic atrophy in the eyes of patients and classified each patient into one of the following four categories:

R + L + : present in both eyes;
R + L − : present in the right but not in the left eye;
R − L + : absent from the right but present in the left eye;
R − L − : absent from both eyes.

In this way, he could examine all four combinations, but could also compare the right eye versus the left eye both the right and the left eye.

For many not so often occurring empirical research situations, extensions of the indices by Scott and Cohen are presented in Gwet (2014) and Shoukri (2011).

3.6 Different Views on Agreement

In the situation where the units are coded, more than twice different views on agreement can be followed, as was discussed at the beginning of this chapter. Pairwise agreement (the mean agreement over all pairs of raters is used) is preferred, but simultaneous agreement (all raters agree in assigning a unit to the same category) or majority agreement (at least k of the m raters agree in assigning a unit to the same category) is possible. A disadvantage of majority agreement is that it is relatively easy to come to complete agreement here in assigning a specific unit. On the other hand, simultaneous agreement sets the highest demands. When just one rater codes a unit into another category than the other raters do, there is no agreement with respect to this unit. For this reason, the estimate of the index will decrease relatively faster than according to the other views when the numbers of disagreements per unit increase. This probably even will go faster when the number of raters increases. Popping (2010) investigated the behavior of indices if these three types for a small data set in which disagreement increased gradually. The results are in Fig. 3.1.

Figure 3.1 contains data on 30 units that are each coded three times into one of three categories. For majority agreement, it is necessary that at least two out of the three ratings are identical. Each rater used each category 10 times. Therefore, the chance expected agreement (to be discussed in detail later) is for each method in all situations identical; i.e., for pairwise, simultaneous and majority agreement respectively 0.33, 0.04, and 0.26. Differences in the outcome of the index are

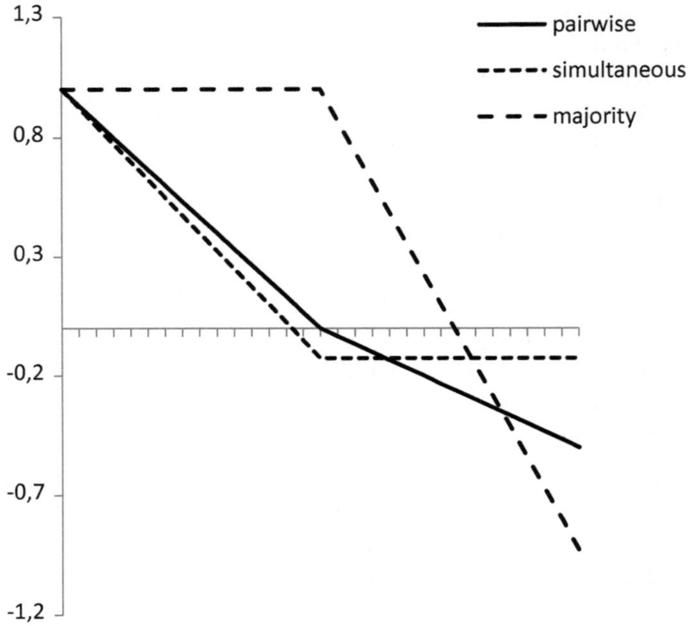

Fig. 3.1 Outcomes pairwise, similarity, and majority agreement. *Source* Popping (2010)

caused by the observed agreement. At first, there was no disagreement in coding, for this reason all indices will take the value 1. In the second step, one rater had once assigned a unit to category 2 instead of category 1 and once to category 1 instead of category 2. So, in each step there will be two more codings that are not in agreement. In the situations of pairwise and simultaneous agreement, the index will now take the value 0.93. Note, however, that observed agreement is different for the three views. In the pairwise comparison, it is 0.96, in the simultaneous comparison 0.93, and in the case of majority agreement, the value is still 1. In this situation, the agreement index can take a high value very soon. In the worst case, all raters disagree in assigning the categories. They all use a different category. Now, the index for pairwise agreement is −0.50, for similarity agreement −0.04, and for majority agreement −0.33.

As long as at least 2 out of the 3 codings agree the kappa for majority agreement takes the value 1. Here, this is in half of the situations. In the last one of these situations, it holds for all units that only one pair agrees. Now, the index for pairwise agreement will take the value zero. As soon as the codings are no longer in agreement for at least one unit, the observed agreement for the simultaneous version of the index becomes zero, and therefore, the index takes a constant value −0.04. This also happens in half of the situations, but here we have a result nobody should be satisfied with. Only the index for mean pairwise agreement steadily decreases from 1 to −0.50. This might be considered as a plea for the index based on a pairwise comparison of the raters.

Mean pairwise agreement looks preferable when one looks at the development of the outcome values in a situation where agreement decreases in a systematic way. This type of agreement is in line with the fraction of agreements. In the situation of majority agreement, the perfect situation is easily reached, even when in reality there is no perfect agreement. For simultaneous agreement, it is reverse. It looks as if there is no agreement at all; this already occurs when all but one raters agree in the assignment. Further on requirements to agreement indices will be mentioned; one of these is ease of interpretation. The interpretation of these two types of indices is hard. Hereafter, only pairwise agreement is considered.

An index for simultaneous agreement has been proposed by Ross (1977), Hubert (1977), and Conger (1980); an index for majority agreement was proposed by Conger (1980) and by Craig (1981). None of the authors compared their method to any other method.

3.7 Selection and Training of Raters

It is clear now that disagreement among raters can be due to all kinds of complications that occur around the coding process. We have seen that raters, categories, and units are relevant for this process. Raters finally have to do the job. At a given moment, the possible problems around units and categories are solved. Now, the raters come in, and possible problems with raters are not solved at once. This makes that selection and training of raters becomes very important. It is during the training that the investigator tries to get the raters on one line. Part of the training can be the learning how to use computer-assisted coding, which structures the coding process. This is not discussed here, but it asks for additional qualities the raters should get. Think of corrections that should be made after incorrect coding. In training programs, attention should be given to the following:

- Raters should be aware of the goal of the investigation. Now, they will understand why the coding is performed as it is performed. On the other hand, one can also take the position that raters should not be aware of the goal. This is to make that, for example in situations of doubt, they will not code in line with the research question;
- Raters should exactly know the meaning of the categories used, see the demands formulated by Holsti that were mentioned before and what follows;
- Raters should be trained in identifying the target behavior. In case they have to code different categories at once, they should be trained in which category to look first at;
- In case there are reasons to assign different categories to a unit, they should know which category is to be preferred in which situation and why.

Clarity of categories, another big source of errors, means that the rater must understand what is meant by the category, but also that this rater has to be qualified to decide whether a unit fits in that category or not. Categories containing 'mostly this' or 'mostly that,' for example, cause a lot of problems.

Ambiguity is very complex. For the moment, it is enough to indicate that by saying 'it drives me crazy' people usually do not mean that they will become crazy but that they are confused. Also, in observation studies behaviors might be found from which one can ask whether it is the intended behavior or not. Furthermore, it would be helpful when raters are trained to motivate their assignments. In case of doubt, they should have the possibility to reconsider the coding.

If systematic disagreement did rise because one rater used the criteria for coding in another way than the other rater, this can be found back in the 'agreement' between different categories, and this intercategory agreement will be high. For the situation all raters deviated, Krippendorff (2008) contains some indices that might be used.

For each coding tasks, one needs a coding instrument. The selection and training of the raters is part of this instrument. A good investigator assures that all raters get a training in which they learn how to look at the units and at certain attributes that might be relevant in the coding task. Very important is that the raters are able to recognize the categories. In many studies, this is a very difficult task, which is often underestimated. All raters should be equally trained. This is especially so when there is no good or wrong assignment. Usually to be coded is whether a certain issue is available (according to the rater) in the unit or not. Even a rater who has the expert role needs certain instructions with regard to the coding. In nearly all studies, the raters are relatively interchangeable. Sometimes a rater acts as a standard; this is certainly true in the training.

One important distinction a rater has to learn is the difference between observation and interpretation. In an observation study, for example, the rater has to code that a specific behavior takes place, but not why such behavior takes place.

If a rater does not meet a certain standard during the training, it is possible that the rater has to be taken of the coding task. This standard is based on a comparison to a standard or to other raters. Some studies will employ multiple coders working independently on the same date. This minimizes the chance of errors from coding and increases the reliability of data.

Burton (1981: 956) observed that 'scorers may have a unique preference for some category … [It is] very difficult to train scorers to abandon such preferences—[our] attempts have generally lead to overcompensation.' If possible, such raters should be replaced by better ones. Knowing that it might be very hard to find good people who can act as a rater, it might also be that still more training might be needed, where some parts maybe should be performed in another way. (Burton allows mild rater preferences, for that reason she claims that the index according to Scott cannot be used any more, the marginal distributions for all raters are not equal anymore.)

The process of rater training is inextricably linked with the process of codebook development, at least for the original development of a codebook. Clear guidelines for the raters are highly relevant. The process may be long and arduous. What follows would be the extreme scenario, including all possible points for codebook revision, in the case of the first development of a coding scheme (Neuendorf 2002: 134).

- Write codebook, with variable selection following guidelines and variable measurement following guidelines;
- Rater training, with discussion;
- Raters practice code together, engaging in consensus-building discussion;
- Possible codebook revisions;
- Rater training on revisions;
- Raters practice code independently on a number of units representing the variety of the population;
- Raters discuss results of independent practice coding;
- Possible codebook revisions;
- Rater training on revisions;
- Raters code pilot subsample for reliability purposes;
- Researcher checks reliabilities;
- Possible codebook revisions;
- Rater training on revisions;
- Final, independent coding (including final reliability checks);
- Rater 'debriefing,' which asks raters to analyze their experiences (may be written).

During the training, agreement might be computed in order to get data on the progress in skills as shown by the future raters. Sometimes their assignments can be compared to the ones by someone whose assignments are considered as a standard. This person might be a senior, anyway someone who makes a correct assignment. An example is the physician who rates patients together with students who are in training to become such a physician. In this situation, the chance expected agreement is often only based on the marginal distribution as provided by this standard.

In case such a standard is not available, the assignments by the rater under training might be compared to the assignments by other raters who are already trained. Now, chance expected agreement is based on the marginal distribution by these trained raters.

A point of discussion is the data to be used in the training. In particular, two views are followed to decide on these test data. Assume the set is not the complete data set:

- The test data might be a random subsample of the total data set;
- The test data might be a set containing all difficult decision moments that can (according to the investigator) occur in the data.

I am not aware of studies telling which view will lead to better-trained raters. I prefer the first view. Complex situations will be included in the subsample based on their relative occurrence.

When coding is performed by human coders, these persons make several decisions based on coding instructions provided by the investigator. Unfortunately, these instructions are usually not reported, leaving details of the coding process closed to public scrutiny. Such details are inevitably instructive when reported, however. Hak and Bernts (1996) analyzed transcripts of discussions between coders and showed how these often led to coding decisions that are not fully justified by formal coding instructions. They argue that coders often base their coding decisions on criteria that emerge during the training and in discussion with other coders. These criteria typically go beyond the formal, written coding instructions. Therefore, they believe, high levels of interrater reliability may often represent little more than a training artifact. Coding instructions must form the basis of all coding decisions. If coding decisions are problematic, better coding instructions are called for. Instructions will differ depending on the instrumental or representational approach and on the nature of ambiguity in texts.

Several questions concerning the qualities of the raters still are not be answered here, this despite their relevance. After performing a coding task resulting in high agreement, the raters are said to be interchangeable. Are we referring to the raters participating in the study now? What about the raters who received the same training, but did not participate for whatever reason? And, what about raters who will receive the same training or maybe a similar training by others. Will 'interchangeable' raters perform as well when another sample of units from the same population is concerned? What if the categories are again the same and the units are of the same type but from a different population?

3.8 Interval or Ordinal Level of Measurement

The discussion so far concerns almost exclusively data at the nominal level of measurement. An investigator might have ordered categories, however, data at the interval or ordinal level of measurement. In a survey, such data are not problematic. Commonly, the respondent gives an opinion concerning an attitude, and one respondent might be more severe than another with regard to a very concrete event. The mean score or the dominant score gives a good impression of how the attitude is rated by respondents. In agreement studies, this is actually different. Raters having the same skills and using identical criteria are supposed to arrive at an identical coding, no matter the difficulty of distinguishing between two sequential categories. In case one rater is more severe than another differences in assigning result. Differences between raters can be small or large. The categories used are close to each other (or example *good* and *very good*), this indicates little disagreement. The categories can also be far from each other (say *good* and *bad*), which refers to complete disagreement. This second situation is worse than the first one. The differences in assignment are considered by using weights. Differences are 'rewarded,' and no complete agreement is found between the raters, but a portion of agreement. Only when the disagreements are maximal, the decision is that there is no agreement. An example of ordered categories is:

1. certain multiple sclerosis;
2. probable multiple sclerosis;
3. possible multiple sclerosis [odds 50: 50];
4. doubtful, unlikely, or definitely not multiple sclerosis.

In case an investigator wants this, the use of weights now is a simple extension to our type of indices to take different assignments to these levels of measurement into account and so to take this kind of problems into account. One must be careful; however, weights can be arbitrarily defined. This is not true for the so-called linear and quadratic weights.

In case data are at least ordered, one might distinguish between two different sources of disagreement. Raters may differ in the definition of the attribute to be rated itself, but also in their definitions of specific rating levels of categories. In both cases, the meaning of the categories is not clear.

A matrix w is used for weights, w_{ij} is the weight for the situation in which one rating of a unit is to category i and the other rating is to category j. It must always be true that $0 \leq w_{ij} \leq 1$. In the standard situation, one uses $w_{ii} = 1$ to indicate perfect agreement and $w_{ij} = 0$ (where $i \neq j$) in case there is no agreement. The algorithm $w_{ij} = 1 - |i - j|/(c - 1)$ is used to indicate a linear weighted relation (Goodman and Kruskal 1954: 758), and $w_{ij} = 1 - (i - j)^2/(c - 1)^2$ refers to the squared weighted relation (Fleiss and Cohen 1973). In both cases, it holds that the greater the disagreement, the lower the weight. In practice, these algorithms are used when the data are on an ordinal or interval level of measurement. Weights are not used that often as a theoretical basis is needed.

In case weights are used, the outcomes found for the γ_3 index might be dependent on the number of categories used. In case one of the algorithms mentioned above is used, certainly the one for the squared relation, the more the categories imply that the difference between sequential weights becomes smaller, weights close to the diagonal will not differ that much from the weights on the diagonal (Brenner and Kliebsch 1996).

Yang and Chen (1978) emphasize that not only $0 \leq w_{ij} \leq 1 = w_{ii}$ must hold, it also must be true that the matrix of weights is symmetric: $w_{ij} = w_{ji}$. In case this condition is not fulfilled, it is always possible based on the weights chosen to get as an outcome that $\gamma_3 = 0$, no matter how the available data do look like. Schouten (1982) showed that using $w_{ij} = w_{ji} = 1$ is a way to combine the categories i and j into one category. Own weights are not recommended, because often no clear rationale is available for the choices made. One other issue is that the matrix of weights has to be symmetric.

The squared weights version of the index is most popular, because this version can be interpreted as an intraclass correlation coefficient (Fleiss and Cohen 1973). If, in addition, the marginal distributions are identical, then weighted kappa is equivalent to the product-moment correlation coefficient. In general, the weighted index tends to behave as a measure of association instead as an agreement coefficient (Graham and Jackson 1993).

The intraclass correlation coefficient (ICC) is used for assessing reliability between multiple raters for data with or without replications. Definitions of different versions of ICCs depend on the assumptions of specific ANOVA models. The parameter estimator for the ICC is usually based on the method of moments with the underlying assumed ANOVA model.

So, the ICC is usually used to compute reliability in case the level of measurement of the data is at least interval. Such data might be approached by using weights. The ICC can be presented as a measure of the degree of homogeneity of the raters relative to the total variability in the interval scale. It compares differences between the scores of each rater and a 'pooled' arithmetic mean, computed over the scores of all raters. As interval data are not the scope of this text, we will only present a start. Assume each unit is rated by a different set of m raters, randomly selected from a larger population of raters. For this situation, a first model applies: ICC = (MSB − MSW)/[MSB + $(m - 1)$ MSW], where MSB denotes the estimate of the variance between raters and MSW the estimate of the within-unit variance. The ICC ranges from $-1/(m - 1)$ to 1.0. It is 1.0 when the within-units variance is zero and the mean square between units is greater than zero. A within variance of zero indicates identical codings for a unit (i.e., all raters agree on the coding of that unit) and hence is consistent with a reliability of 1, denoting perfect agreement. A negative intraclass correlation is usually interpreted as no reliability. In case there is some error, the part 1—ICC is interpreted as the percentage of variance due to the disagreement among the raters. Fleiss and Cohen (1973) have shown that there is 'equivalence' between the squared weighted kappa according to Cohen and the ICC when 'the systematic variability between raters is included as a component of variability.' The difference between the indices depends on the difference between the arithmetic and the geometric mean. These means usually are close to each other.

In case the level of measurement is nominal or dichotomous, other variations of the ICC might be used. The requirements of identical situation and of independence between the codings become difficult. A rater might remember what coding was used the previous time. For equivalences between the kappa coefficient and the ICC for several different research designs, see among others Bartko (1966), Fleiss, and Cohen (1973) or Shrout and Fleiss (1979).

The notion of agreement sometimes appears in the form of internal consistency in scale development. When a set of items are asked to a sample of respondents (units) in a study in order to measure a specific construct, the scale developer expects all of the items to show (internal) consistency toward the measurement of a unique latent construct. High internal consistency is an indication of a high degree of agreement among the items with respect to the construct associated with the respondents. Well-known measures of internal consistency are Cronbach's alpha coefficient (Cronbach 1951) and Mokken's coefficient (Mokken 1971).

References

A. Agresti, Modelling patterns of agreement and disagreement. Stat. Methods Med. Res. **1**(2), 201–218 (1992)

M. Aickin, Maximum likelihood estimation of agreement in the constant predictive probability model, and its relation to Cohen's kappa. Biometrics **46**(2), 293–302 (1990)

R.F. Bales, *Interaction Process Analysis. A Method for the Study of Small Groups* (Addison Wesley, Reading, MA, 1950)

M. Banerjee, M. Capozzoli, L. McSweeney, D. Sinha, Beyond kappa: a review of interrater agreement measures. Can. J. Stat. **27**(1), 3–23 (1999)

J.J. Bartko, The intraclass correlation coefficient as a measure of reliability. Psychol. Rep. **19**(1), 3–11 (1966)

J.J. Bartko, W.T. Carpenter, Methods and theory of reliability. J. Nerv. Mental Dis. **163**(5), 307–317 (1976)

E.M. Bennett, R.L. Blomquist, A.C. Goldstein, Response stability in limited response questioning. Publ. Opin. Q. **18**(2), 218–223 (1954)

D.A. Bloch, H.C. Kraemer, 2×2 kappa coefficients: measures of agreement or association. Biometrics **45**(1), 269–287 (1989)

R.L. Brennan, D.J. Prediger, Coefficient kappa: some uses, misuses, and alternatives. Educ. Psychol. Meas. **41**(4), 687–699 (1981)

H. Brenner, U. Kliebsch, Dependence of weighted kappa coefficients on the number of categories. Epidemiology **7**(2), 199–202 (1996)

N.W. Burton, Estimating scorer agreement for nominal categorization systems. Educ. Psychol. Meas. **41**(4), 953–962 (1981)

A.B. Cantor, Sample size calculations for Cohen's kappa. Psychol. Methods **1**(2), 150–153 (1996)

D.V. Cicchetti, A.R. Feinstein, High agreement but low kappa, II: resolving the paradoxes. J. Clin. Epidemiol. **43**(6), 551–558 (1990)

J. Cohen, A coefficient of agreement for nominal scales. Educ. Psychol. Meas. **20**(1), 37–46 (1960)

J. Cohen, Weighted kappa. Nominal scale agreement with provision for scaled disagreement or partial credit. Psychol. Bull. **70**(4), 213–220 (1968)

A.J. Conger, Integration and generalization of kappas for multiple raters. Psychol. Bull. **88**(2), 322–328 (1980)

P.E. Converse, The nature of belief systems in mass publics. Ideology and discontent, in *Ideology and Discontent*, ed. by D.E. Apter (The Free Press, New York, 1964), pp. 206–261

R.T. Craig, Generalization of Scott's index of intercoder agreement. Public Opin. Q. **45**(2), 260–264 (1981)

K.S. Crittenden, Actual and reconstructed coding procedsure, in *Academic Janus*, ed. by R. McGee (Jossey Bass, San Francisco, 1971), pp. 228–246

K.S. Crittenden, R.J. Hill, Coding reliability and validity of interview data. Am. Sociol. Rev. **36** (6), 1073–1080 (1971)

L.J. Cronbach, Coefficient alpha and the internal structure of tests. Psychometrika **16**(3), 297–334 (1951)

J.N. Darroch, P.I. McCloud, Category distinguishability and observer agreement. Aust. NZ. J. Stat. **28**(3), 371–388 (1986)

H.C. De Vet, L.B. Mokkink, C.B. Terwee, O.S. Hoekstra, D.L. Knol, Clinicians are right not to like Cohen's κ. Br. Med. J. **12**(346), f2125 (2013)

W. Dijkstra, T. Taris, Measuring the agreement between sequences. Sociol. Methods Res. **24**(2), 214–231 (1995)

P.F. Duggan, Time to abolish 'gold standard'. Br. Med. J. **304**(6811), 1568–1569 (1992)

G.C. Feng, Underlying determinants driving agreement among coders. Qual. Quant. **47**(5), 2983–2997 (2013)

V.F. Flack, A.A. Afifi, P.A. Lachenbruch, H.J.A. Schouten, Sample size determinations for the two rater kappa statistic. Psychometrika **53**(3), 321–325 (1988)

N.A. Flanders, Estimating reliability, in *Interaction Analysis: Theory, Research and Applications*, ed. by E.J. Amidon, J.B. Hough (Addison-Wesley, Reading, Mass., 1967), pp. 161–166

J.L. Fleiss, Estimating the accuracy of dichotomous judgements. Psychometrika **30**(4), 469–479 (1965)

J.L. Fleiss, Measuring nominal scale agreement among many raters. Psychol. Bull. **76**(5), 378–382 (1971)

J.L. Fleiss, Measuring agreement between two judges on the presence or absence of a trait. Biometrics **31**(3), 651–659 (1975)

J.L. Fleiss, Response to Krippendorff. Biometrics **34**(1), 144 (1978)

J.L. Fleiss, J. Cohen, The equivalence of weighted kappa and the intraclass correlation coefficient as measures of reliability. Educ. Psychol. Meas. **33**(3), 613–619 (1973)

S.L. Foster, J.D. Cone, Design and use of direct observation procedures, in *Handbook of Behavioral Assessment*, ed. by A.R. Ciminero, K.S. Calhoun, H.E. Adams (Wiley, New York, 1986), pp. 253–324

J. Galtung, *Theory and Methods of Social Research* (Allen & Unwin, London, 1967)

J. Galtung, Measurement of agreement, in *Papers on Methodology. Theory and Methods of Social Research. Volume II*, ed. by J. Galtung (Christian Eijlers, Copenhagen, 1979), pp. 82–135

L.A. Goodman, W.H. Kruskal, Measures of association for cross classifications. J. Am. Stat. Assoc. **49**(268), 732–764 (1954)

P. Graham, R. Jackson, The analysis of ordinal agreement data: beyond weighted kappa. J. Clin. Epidemiol. **46**(9), 1055–1062 (1993)

K.L. Gwet, *Handbook of inter-rater reliability. The definitive guide to measuring the extent of agreement among raters* (Advanced Analytics LLC, Gaithersburg, USA, 2014)

K.L. Gwet, Computing inter-rater reliability and its variance in the presence of high agreement. Br. J. Math. Stat. Psychol. **61**(1), 29–48 (2008)

T. Hak, T. Bernts, Coder training: theoretical training or practical socialization? Qual. Sociol. **19** (2), 235–257 (1996)

D.P. Hartmann, Considerations in the choice of interobserver reliability estimates. J. Appl. Behav. Anal. **10**(1), 103–116 (1977)

A.R. Hollenbeck, in Problems of reliability in observational research, in *Observing Behavior, Volume 2*, ed. by G.P. Sacker (University Park Press, London, 1978), pp. 79–98

J.W. Holley, J.P. Guilford, A note on the G-index of agreement. Educ. Psychol. Meas. **24**(4), 749–753 (1964)

O.R. Holsti, *Content analysis for the social sciences and humanities* (Addison Wesley, London, 1969)

L.M. Hsu, R. Field, Interrater agreement measures: comments on kappan, Cohen's kappa, Scott's π and Aickin's α. Underst. Stat. **2**(3), 205–219 (2003)

L.J. Hubert, Kappa revisited. Psychol. Bull. **84**(2), 289–297 (1977)

S. Janson, J. Vegelius, On the generalization of the G index and the phi coefficient to nominal scales. Multivar. Behav. Res. **14**(2), 255–269 (1979)

A.E. Kazdin, Artifact, bias, and complexity of assessment: the ABC's of reliability. J. Appl. Behav. Anal. **10**(1), 141–150 (1977)

R.N. Kent, K.D. O'Leary, A. Dietz, C. Diament, Comparison of observational recordings in vivo, via mirror, and via television. J. Appl. Behav. Anal. **12**(4), 517–522 (1979)

K. Krippendorff, Reliability of binary attribute data. Biometrics **34**(1), 142–144 (1978)

K. Krippendorff, Association, agreement, and equity. Qual. Quant. **21**(1), 109–123 (1987)

K. Krippendorff, Reliability in content analysis: some common misconceptions and recommendations. Hum. Commun. Res. **30**(3), 411–433 (2004)

K. Krippendorff, Systematic and random disagreement and the reliability of nominal data. Commun. Methods Meas. **2**(4), 323–338 (2008)

K. Krippendorff, Agreement and information in the reliability of coding. Commun. Methods Meas. **5**(2), 93–112

S. Lacy, D. Riffe, Sampling error and selecting intercoder reliability samples for nominal content categories. J. Mass Commun. Q. **73**(4), 963–973 (1996)

J.R. Landis, G.G. Koch, A review of statistical methods in the analysis of data arising from observer reliability studies. Part 2. Stat. Neerl. **29**(2), 151–161 (1975)

J.R. Landis, G.G. Koch, The measurement of observer agreement for categorical data. Biometrics **33**(1), 159–174 (1977)

R.J. Light, Measures of response agreement for qualitative data: some generalizations and alternatives. Psychol. Bull. **76**(5), 365–377 (1971)

A. Linderman, Computer content analysis and manual coding techniques: A comparative analysis. in *Theory, Method, and Practice in Computer Content Analysis*, ed. by M.D. West (Ablex, Westport, CT, 2001), pp. 97–109

A.E. Maxwell, Coefficients of agreement between observers and their interpretation. Br. J. Psychiatry **130**(1), 79–83 (1977)

S. Mikhaylov, M. Laver, K. Benoit, Coder reliability and misclassification in the human coding of party manifestos. Polit. Anal. **20**(1), 78–91 (2012)

R.J. Mokken, *A Theory and Procedure of Scale Analysis: With Applications in Political Research* (Mouton,The Hague, 1971)

A.C. Montgomery, K.S. Crittenden, Improving coding reliability for open-ended questions. Public Opin. Q. **41**(2), 235–243 (1977)

K.A. Neuendorf, *The Content Analysis Guidebook* (Sage, Thousand Oaks, CA, 2002)

A.M. Noda, H.C. Kraemer, J.A. Yesavage, V.S. Periyakoil, How many raters are needed for a reliable diagnosis? Int. J. Methods Psychiatr. Res. **10**(3), 119–125 (2006)

R. Popping, "Het indelen naar werkoriëntaties" [Classifying job orientations], in *Modellen in de sociologie*, ed. by S. Lindenberg, F.N. Stokman (Van Loghum Slaterus, Deventer, 1983), pp. 233–247

R. Popping, Some views on agreement to be used in content analysis studies. Qual. Quant. **44**(6), 1067–1078 (2010)

R. Popping, On agreement indices for nominal data, in *Sociometric Research, Vol. I*, ed. by W.E. Saris, I.N. Gallhofer (McMillan, London, 1988), pp. 90–105

R. Popping, Analyzing open-ended questions by means of text analysis procedures. Bull. Méthodologie Sociol. **128**, 23–39 (2015)

R. Popping, C.W. Roberts, Coding issues in semantic text analysis. Field Methods **21**(3), 244–264 (2009)

D. Quade, Nonparametric partial correlation, in *Measurement in the Social Sciences. Theories and Strategies*, ed. by H.M. Blalock (MacMillan, London, 1974), pp. 369–398

E. Rogot, I.D. Goldberg, A proposed index for measuring agreement in test-retest studies. J. Chron. Dis. **19**(9), 991–1006 (1966)

R.G. Romanczyk, R.N. Kent, C. Diament, K.D. O'Leary, Measuring the reliability of observational data: a reactive process. J. Appl. Behav. Anal. **6**(1), 175–184 (1973)

D.C. Ross, Testing patterned hypothesis in multi-way contingency tables using weighted kappa and weighted chi square. Educ. Psychol. Meas. **37**(2), 291–307 (1977)

H.J.A. Schouten, Measuring pairwise agreement among many observers. II. Some improvements and additions. Biometrical J. **24**(5), 431–435 (1982)

H.J.A. Schouten, Estimating kappa from binocular data and comparing marginal probabilities. Stat. Med. **12**(23), 2207–2217 (1993)

W.A. Scott, Reliability of content analysis: the case of nominal scale coding. Public Opin. Q. **19** (3), 321–325 (1955)

M.M. Shoukri, *Measures of Interobserver Agreement and Reliability* (CRC Press, Boca Raton, Fl, 2011)

P.E. Shrout, J.L. Fleiss, Intraclass correlations: uses in assessing rater reliability. Psychol. Bull. **80** (2), 420–428 (1979)

P. Simon, Including omission mistakes in the calculation of Cohen's kappa and an analysis of the coefficient's paradox features. Educ. Psychol. Meas. **66**(5), 765–777 (2006)

J. Spanjer, B. Krol, R. Popping, J.W. Groothoff, S. Brouwer, Disability assessment interview: the role of concrete and detailed information on functioning besides medical history taking. J. Rehabil. Med. **41**(4), 267–272 (2009)

A. Stuart, A test of homogeneity of marginal distributions in a two-way classification. Biometrika **42**(3/4), 412–416 (1955)
H.K. Suen, Agreement, reliability, accuracy, and validity: toward a clarification. Behav. Assess. **10**(4), 343–366 (1988)
J.S. Uebersax, Diversity of decision-making models and the measurement of interrater agreement. Psychol. Bull. **101**(1), 140–146 (1987)
J.S. Uebersax, Statistical modeling of expert ratings on medical treatment appropriateness. J. Am. Stat. Assoc. **88**(2), 421–427 (1993)
U.N. Umesh, R.A. Peterson, M.H. Sauber, Interjudge agreement and the maximum value of kappa. Educ. Psychol. Meas. **49**(4), 835–850 (1989)
S. Vanbelle, A. Albert, Agreement between an isolated rater and a group of raters. Stat. Neerl. **63** (1), 82–100 (2009)
J. Vegelius, S. Janson, Criteria for symmetric measures of association for nominal data. Qual. Quant. **16**(4), 243–250 (1982)
J. Vegelius, On the utility of the E-correlation coefficient concept in psychological research. Educ. Psychol. Meas. **38**(3), 605–611 (1978)
A. Von Eye, Alternatives to Cohen's κ. Eur. Psychol. **11**(1), 12–24 (2006)
H.F. Weisberg, Models of statistical relationship. Am. Polit. Sci. Rev. **68**(4), 1638–1655(1974)
J.L. Woodward, R. Franzen, A study on coding reliability. Public Opin. Q. **12**(2), 253–257 (1948)
G.L. Yang, M.K. Chen, A note on weighted kappa. Soc. Econ. Plann. Sci. **12**(5), 293–294 (1978)
R. Zwick, Another look at interrater agreement. Psychol. Bull. **103**(3), 374–378 (1988)

Part II
Formulas

Chapter 4
Indices

This chapter contains the agreement indices that fulfill the requirements that have been discussed before. The set of coding categories to be used might be available when the coding task starts, and this is the common situation. This set is identical for each rater. This situation is referred to as the a posteriori method of coding. First, this situation is considered. It might be however that the raters also have to develop a set of categories themselves. Now each category in the set might have a different content for each rater, and also the number of categories used can be different. This is the a priori method of coding. This research situation is not found that often, it is discussed shortly. After that, indices are listed that do not fulfill the requirements.

In Chap. 3, reference was made to situations in which γ_3-type indices might be problematic, especially in situations of prevalence and bias. The discussion was almost exclusively on κ-type indices, but the problems might also concern π-type indices. Possible solutions are discussed at the end of the part on measures to be used in the a posteriori situation.

In the examples that will be presented, two digits will be shown, although this might sometimes be more for the computation of chance expected agreement of constituting parts of it. This is done to make visible the differences in outcome when the view by Scott or by Cohen is followed. Examples of computations are in terms of frequencies. In this way, one is closest to the original data as presented in a table or a data matrix.

4.1 A Posteriori Method of Coding

The common situation is that the categories to be used are known in advance and are identical for all raters. Agreement is always computed for one item at a time. The data for an agreement study with respect to this one item can be available in two ways. First, there is an agreement table in which the assignments by two raters

R. Popping, *Introduction to Interrater Agreement for Nominal Data*,
https://doi.org/10.1007/978-3-030-11671-2_4

are compared. This situation will be explained first. Next, the other way is considered. Here, one starts from a data matrix. Such a matrix is generally used when there are more than two raters to be compared. Usually, the data matrix is one having in the rows information regarding the units and in the column information regarding the raters. In the cell, one finds the number of the category the corresponding rater assigned to the unit. Another form for the matrix, however, is also possible. This one is sometimes found when the raters themselves are not relevant.

First indices for comparing assignments by two raters will be introduced. With respect to the computation of chance expected agreement, each time indices based on both the view by Scott and the one by Cohen will be presented. This will be announced. The announcement is followed by a reference to the core formulas that are used to compute observed and expected agreement. This should be helpful. As the formulas for observed agreement use usually identical when either Scott or Cohen is followed, this formula is usually mentioned only once.

It is possible that in an analysis the assignments by more than two raters, or more than two ratings, are to be compared. One rater might be a standard (accuracy) or a new rater who is to be added to an existing group of raters. An investigator might be interested in the assignments to one specific category (because it is extremely interesting or very hard to recognize correctly). Indices for all these situations will be discussed. At the end of this part, the state of art with respect to the handling of skewed distributions will be discussed.

4.1.1 Two Raters

In an agreement table, the assignments by two raters are compared. Each rater used the same set of categories; therefore, the table is a squared one. Let us assume c categories have been used. An example of such a table is Table 1.1, where the use of reality claims is compared. In Fig. 4.1 follows a general presentation of the table. In general, one might say f_{ij} denotes the number of assignments to category i by the first rater and to category j by the second rater. The number of assignments by rater 1 to category i is $f_{i(1)}$, and the number of assignments by rater 2 to category j is $f_{j(2)}$. The total number of units that has been assigned to a category is N.

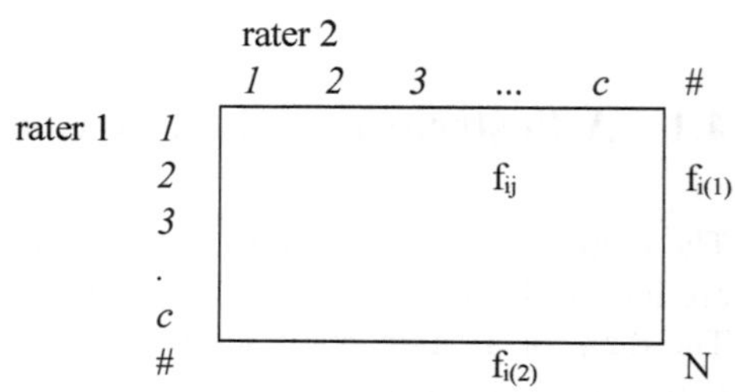

Fig. 4.1 Agreement table containing the assignments by two raters using c categories

All these frequencies f_{ij}, $f_{i(1)}$, and $f_{j(2)}$ can be written in terms of proportions p_{ij}, $p_{i(1)}$, and $p_{j(2)}$. These are found by dividing the frequency by the number of units N. Proportions are preferred, because in that situation there is a clear lower and upper bound.

Scott approach, core formulas (4.2) *and* (4.3)

Starting from equally skilled raters, one would use Scott's π. The index is as the γ_3-index above:

$$\pi = (P_o - P_e)/(1 - P_e), \tag{4.1}$$

where

$$P_o = \sum_{i=1}^{c} f_{ii}/N = \sum_{i=1}^{c} p_{ii}, \tag{4.2}$$

$$P_e = \sum_{i=1}^{c} f_i^2/N^2 = \sum_{i=1}^{c} p_i^2. \tag{4.3}$$

with

$$f_i = \frac{f_{i(1)} + f_{i(2)}}{2}, \quad p_i = \frac{p_{i(1)} + p_{i(2)}}{2}, \tag{4.4}$$

as the mean frequency or proportion per category over all raters is needed.

Table 4.1 contains the same data as does Table 1.1, and only the marginal distribution is equal for both raters based on formula (4.4). The categories refer to the types of reality claims that are found in the sentences that are coded. This is direct observable information, the ideographic behavior paradigm should be followed, and therefore the Scott approach is used. Following (4.2) we compute: $P_o = (217 + 120 + 84 + 8 + 0)/487 = 0.88$ and following (4.3): $P_e = (239.5^2 + 136^2 + 86.5^2 + 9.5^2 + 15.5^2)/487^2 = 0.3528$. From formula (4.1), we see that $\pi = (0.88 - 0.3528)/(1 - 0.3528) = 0.82$. Following Landis and Koch (1977), this result is 'almost perfect.' The investigators have proceeded their analyses using the assignment by one of the raters.

There is another problem with the data in Table 4.1. The category 'not coded' is included. This category contains the units that according to one rater contains a reality claim and according to the other rater it does not. Units that do not contain such a reality claim according to both raters are not included in the table (cell [5, 5] is empty). In case these units would have been included, the distribution would have been very skewed and the frequency in the cell would have been 4787. This is a choice by the investigators. In case only units would have been included that are considered as reality claims by both raters, one would find $\pi = 0.9054 = (0.94 - 0.3744)/(1 - 0.3744)$ given $N = 456$. This result is much better than the one that was found when the 'not coded' category was still included.

Table 4.1 Two raters' assignments of four reality claims, total distribution of categories based on alignments by all raters

		Rater 2					
		Possible	Impossible	Inevitable	Contingent	Not coded	#
Rater 1	Possible	217	24	0	0	15	239.5
	Impossible	2	120	0	0	2	136.0
	Inevitable	0	0	84	0	2	86.5
	Contingent	0	1	0	8	0	9.5
	Not coded	4	3	3	2	0	15.5
	#	239.5	136.0	86.5	9.5	15.5	487

Cohen approach, core formulas (4.6) *and* (4.7)

Following the approach suggested by Cohen, one gets:

$$\kappa = (P_o - P_e)/(1 - P_e), \tag{4.5}$$

where P_o is as before

$$P_o = \sum_{i=1}^{c} f_{ii}/N = \sum_{i=1}^{c} p_{ii}, \tag{4.6}$$

which is identical to (4.2). Hereafter, identical formulas will not be repeated anymore. However,

$$P_e = \sum_{i=1}^{c} \frac{f_{i(1)}f_{i(2)}}{N^2} = \sum_{i=1}^{c} p_{i(1)}p_{i(2)}. \tag{4.7}$$

As an example, we use fictitious assignments based on diagnoses by psychiatrists; these are their interpretations of what is going on with the patient. The nomothetic-trait paradigm applies.

Table 4.2 contains diagnostic placements by two psychiatrists. Let us assume these psychiatrists operated in line with the view suggested by Cohen. Now we find: $P_o = (80 + 20 + 20)/200 = 0.6$, and $P_e = (100 \times 120 + 60 \times 40 + 40 \times 40)/(200 \times 200) = 0.4$. The result is $\kappa = (0.6 - 0.4)/(1 - 0.4) = 0.33$. Using the

Table 4.2 Two raters' observed diagnostic placements

		Rater 2			
		Schizophrenic	Manic-depressive	Neurosis	#
Rater 1	Schizophrenic	80	40	0	120
	Manic depressive	0	20	20	40
	Neurosis	20	0	20	40
	#	100	60	40	200

criteria mentioned by Landis and Koch (1977), this would point to a fair agreement. As these are example data, we do not know how an investigator would have acted. My guess would be that the criterion for reliability would demand a higher outcome for the index. At this moment, it looks that the differences between the coders are big. The investigator should find out why this is so.

The outcome found after using the formula according to Scot t is always less or equal to the outcome according to Cohen. This is because the chance expected agreement as Scott computed it is equal or higher than the chance expected agreement according to Cohen. This is shown easily. In order to let this be true, we need for category *i* that

$$\left(p_{i(1)} + p_{i(2)}\right)/2 \geq p_{i(1)} p_{i(2)}. \tag{4.8}$$

This is so. Multiply both sides with 2 and bring the part at the right to the left:

$$p_{i(1)} + p_{i(2)} - 2p_{i(1)}p_{i(2)} = p_{i(1)}\left(1 - p_{i(2)}\right) + p_{i(2)}\left(1 - p_{i(1)}\right). \tag{4.9}$$

This expression always is ≥ 0, because the marginal proportion always has as its lower limit the value 0 and as its upper limit the value 1. A consequence is that the index according to Scott is more conservative than the one according to Cohen, for some an argument to prefer the Scott version.

4.1.2 Weights

In order to give an idea of how weights are used in formulas, they are presented here for the present research situation. After that, the situation that weights are used is ignored, and however, two sections will be excluded in which agreement per category is considered. Here, weights can be used to create a situation in which one has two categories: the one of interest verses all others. In case weights are used, P_o and P_e are defined in a different way than presented before because now also non-diagonal cells in the agreement table will contribute to the mount of agreement found.

Scott approach, core formulas (4.10) *and* (4.11)

In both the Scott and the Cohen approach, the definition of P_o is identical:

$$P_o = \sum_{i=1}^{c} \sum_{j=1}^{c} w_{ij} f_{ij} / N = \sum_{i=1}^{c} \sum_{j=1}^{c} w_{ij} p_{ij}. \tag{4.10}$$

The definition of P_e is again different for Scott and Cohen. For Scott, we find:

$$\begin{aligned} P_e &= \sum_{i=1}^{c}\sum_{j=1}^{c} w_{ij}\frac{f_{i(1)}+f_{i(2)}}{2}\times\frac{f_{j(1)}+f_{j(2)}}{2}/N^2 \\ &= \sum_{i=1}^{c}\sum_{j=1}^{c} w_{ij}\frac{p_{i(1)}+p_{i(2)}}{2}\times\frac{p_{j(1)}+p_{j(2)}}{2}. \end{aligned} \tag{4.11}$$

Cohen approach, core formulas (4.10) *and* (4.12)

For Cohen, the formula is:

$$P_e = \sum_{i=1}^{c}\sum_{j=1}^{c} w_{ij}f_i f_j/N^2 = \sum_{i=1}^{c}\sum_{j=1}^{c} w_{ij}p_i p_j. \tag{4.12}$$

Hereafter, formulas will be presented for both the unweighted and the weighted version of the index that is discussed. This is done because the formula for the unweighted version of the index usually looks much simpler than the weighted version. Remember however that with respect to data at the nominal level it holds that w_{ii} = 1 and w_{ij} = 0 ($i \neq j$), where i and j refer to the sequence number of the category.

Besides from here on only the situation is considered in which there are more than two raters to be compared (unless this seems useless). Each time, observed agreement and chance expected agreement are based on the mean of these types of agreement over all pairs of raters. The situation of two raters now is a special case in the comparison over pairs as there is only one pair of raters.

The index γ_3 can also be presented in terms of disagreement. Say the observed disagreement $Q_o = 1 - P_o$ and the disagreement by chance is $Q_e = 1 - P_e$. Now:

$$\gamma_3 = (Q_e - Q_o)/Q_e = 1 - Q_o/Q_e. \tag{4.13}$$

Here, it is not relevant whether chance disagreement is defined following Scott or following Cohen.

For Cohen, it is important that the raters' marginal distributions can be different. In case such a difference exists, the maximum possible value the index takes cannot be 1. The outcome γ_3 = 1 would be found in case the amount of observed agreement also is 1. The maximum observed agreement is based on the marginal distributions. It is found by summing over all categories the smaller of the two marginal totals (Cohen 1960: 42):

$$P_m = \sum_{k=1}^{c} \min\left(p_{i(1)}, p_{i(2)}\right). \tag{4.14}$$

The maximum possible estimate for κ can be computed by using this outcome, say this is κ_{m}.

$$\kappa_{\mathrm{m}} = (P_{\mathrm{m}} - P_{\mathrm{e}})/(1 - P_{\mathrm{e}}). \tag{4.15}$$

The amount $1 - \kappa_{\mathrm{m}}$ represents the proportion of the possibilities of agreement (chance excluded) which cannot be achieved as a consequence of differing marginals. The latter quantity can serve as an indicator to the investigator of the fuzziness of his category boundaries and may be reduced and his boundaries sharpened by further training of his judges.

The ratio $\kappa/\kappa_{\mathrm{m}}$ then can be used to determine how much of the marginally permitted agreement is present between raters. For the data in Table 4.2, we find the maximum possible amount of observed agreement $P_{\mathrm{m}} = (100 + 40 + 40)/200 = 0.9$. This makes $\kappa_{\mathrm{m}} = (0.9 - 0.4)/(1 - 0.4) = 0.83$. This implies that the amount 0.17 cannot be achieved. It also means that the ratio 0.33/0.83 = 0.40 of the marginally permitted agreement is present between the raters.

4.1.3 Intracategory Agreement for Two Raters

In the situation of intracategory agreement, the π-index from formula (4.1) or the κ-index from formula (4.5) reduces to an index in which only one specific category is considered. In case this is category i, the index becomes π_i or κ_i. Intracategory agreement helps in finding the category that caused difficulties with respect to agreement. This difficulty is made visible by the fact that agreement for such categories is lowest. The method is only useful in case more than two categories are used. The method does not inform about the why of these difficulties, however. The type of agreement might be considered in two ways.

First, the category of interest i is assigned to the unit or not, so the other categories are as it were lumped into one. Based on the data in Table 1.1, one will get now as it where what is in Table 4.3.

Starting from an agreement table:

$$f_{\bar{i}\bar{i}} = \sum_{\substack{k=1\\k\neq i}}^{c} \sum_{\substack{j=1\\j\neq i}}^{c} f_{kj}. \tag{4.16}$$

Table 4.3 Table 1.1 split in sub-tables per category

	Possible			Impossible			Inevitable			Contingent			Not coded		
	+	−	#	+	−	#	+	−	#	+	−	#	+	−	#
+	217	39	256	120	4	124	84	2	86	2	10	12	0	12	12
−	6	225	231	24	335	363	3	398	401	8	467	475	19	456	475
#	223	264	487	148	339	487	87	400	487	10	477	487	19	468	487

+ means reality claim applies; − means reality claim does not apply

This might also be presented in another way, one in which an uncommon way of weighting has been applied. Say we use again the data from Table 1.1 and wants to compare category 1 to the other categories. The matrix of weights now is as in Table 4.4.

Second, the frequency in the category of interest i is based on the mean number of assignments per rater to that category. Here, all categories are used, and therefore, weights can be included.

The indices for observed agreement per pair following Scott are different from those following Cohen, because the marginal distributions play a role now, and these marginal distributions are considered differently. Scott uses the mean assignment to each category; Cohen used the distribution over categories per rater. Taking this difference into account, the final observed agreement is the mean over all pairs. The final chance expected agreement following Cohen is also based on the mean over pairs. When Scott is followed, the chance expected agreement is computed in one step.

In the examples in this section, assignments by two raters are used. In a later section, the assignments by more raters are used, and here, intracategory agreement will be considered again.

Scott approach, category of interest or not, core formulas (4.17) *and* (4.18)

Define $P_{o(i\text{-}gh)}$ as the observed agreement with respect to category i for raters g and h. For the first view based on assignment to the category of interest or not according to Scott, we find (Schouten 1986):

$$P_{o(i)} = (f_{ii} + f_{\bar{i}\bar{i}})/N = 1 - 2(f_i - f_{ii})/N = 1 - 2(p_i - p_{ii}). \tag{4.17}$$

Chance expected agreement for the single category can also be formulated in terms of assigned to category i or not assigned to category i:

$$P_{e(i)} = (f_i/N)^2 + \langle (N - f_i)/N \rangle^2 = p_i^2 + (1 - p_i)^2 = 1 - 2p_i(1 - p_i). \tag{4.18}$$

The outcomes for the separate categories from the data in Table 4.1 are in Table 4.5. An example based on the assignments by more than two raters follows

Table 4.4 Matrix of weights for comparison category of interest or not

		Rater 2				
		Possible	Impossible	Inevitable	Contingent	Not coded
Rater 1	Possible	1	0	0	0	0
	Impossible	0	1	1	1	1
	Inevitable	0	1	1	1	1
	Contingent	0	1	1	1	1
	Not coded	0	1	1	1	1

Table 4.5 Outcomes intracategory agreement for data from Table 4.1 based on category of interest or not based on Scott

Cat.	Label	P_o	P_e	π
1	Possible	0.91	0.5001	0.82
2	Impossible	0.93	0.05975	0.84
3	Inevitable	0.99	0.07079	0.96
4	Contingent	0.99	0.08617	0.84
5	Not coded	0.94	0.09384	−0.03

$N = 487$

later. For the first category 'possible' in Table 4.1, we find $P_{o(i-12)} = 1 - 2 \times (239.5 - 217)/487 = 0.91$ and $P_{e(i)} = 1 - 2 \times (239.5/487) \times (1 - 239.5/487) = 0.5001$. From this, we can compute $\pi = (0.91 - 0.5001)/(1 - 0.5001) = 0.82$.

Again, using the Landis and Koch (1977) suggestion the results for the first four categories are 'almost perfect.' The result for the fifth category, however, is horrible. But this was to be expected, the 'not coded' by both raters, i.e., both raters agreed on the fact that there is no reality claim in the clause, was not counted. Therefore, the last cell in the table is empty, only off-diagonal frequencies are possible.

The computation of π using weights as in Table 4.4 results in the same outcome as found for category 1 above. The same holds in case the index would have been computed for each of the five sub-tables in Table 4.10.

Scott approach, mean assignment to category of interest, core formulas (4.19) *and* (4.20)

For the second view based on the mean assignment to the category of interest according to Scott (Fleiss 1971):

$$P_{o(i)} = f_{ii}/f_i = p_{ii}/p_i. \tag{4.19}$$

When the average use of the category is considered, the chance expected agreement for the single category i according to the Scott model is

$$P_{e(i)} = p_i^2/p_i = p_i. \tag{4.20}$$

The outcomes for the first category of the data in Table 4.1 are $P_o = 217/239.5 = 0.91$ and $P_e = 239.5/487 = 0.4918$. The outcomes for all separate categories are in Table 4.6.

The estimates found for π_i are the same as before. In case formulas (4.17) and (4.18) are entered into the formula, or the formulas (4.19) and (4.20), it is easily seen that one will always get

$$\pi_i = \left(p_{ii} - p_i^2\right) / \left(p_i - p_i^2\right). \tag{4.21}$$

Table 4.6 Outcomes intracategory agreement for data from Table 4.1 based on mean assignment to category based on Scott

Cat.	Label	P_o	P_e	π
1	Possible	0.91	0.4918	0.82
2	Impossible	0.88	0.02793	0.84
3	Inevitable	0.97	0.01776	0.96
4	Contingent	0.84	0.00195	0.84
5	Not coded	0.00	0.00318	−0.03

$N = 487$

However, this is only true in case no weights are used. As the outcomes in both situations are identical, also both times the same interpretation holds. The overall π for two raters can also be computed based on the mean assignment per category:

$$\pi = \sum_{i=1}^{c} p_i(1 - p_i)\pi_i / \sum_{i=1}^{c} p_i(1 - p_i). \tag{4.22}$$

Note that in case one wants to use weights anyway, this is only possible when this second view is followed. Now these weights will be incorporated in the comparison for the pair.

Cohen approach, category of interest or not, core formulas (4.23) *and* (4.24)

Next follow the two views in terms of the Cohen approach. First again the situation where assignment to the category of interest or not is considered (Fleiss 1981: 220):

$$P_{o(i)} = (f_{ii} + f_{\bar{i}\bar{i}})/N = 1 - \left(f_{i.(1)} + f_{i.(2)} - 2f_{ii}\right)/N = 1 - p_{i(1)} - p_{i(2)} + 2p_{ii}. \tag{4.23}$$

In this model, chance expected agreement on category i for raters g and h is found by using:

$$\begin{aligned} P_{e(i)} &= \left(f_{i(1)}f_{i(2)} + f_{\bar{i}(1)}f_{\bar{i}(2)}\right)/N^2 = \left(N - f_{i(1)} - f_{i(2)} + 2f_{i(1)}f_{i(2)}\right)/N \\ &= 1 - p_{i(1)} - p_{i(2)} + 2p_{i(1)}p_{i(2)}. \end{aligned} \tag{4.24}$$

Table 4.7 Outcomes intracategory agreement based for data from Table 4.2 on category of interest or not based on Cohen

Cat.	Label	P_o	P_e	κ
1	Schizophrenic	0.70	0.5000	0.40
2	Manic depressive	0.70	0.6200	0.21
3	Neurosis	0.80	0.6800	0.38

$N = 200$

For the first category of the data in Table 4.2, we find $P_o = 1 - (100 + 120 - 2 \times 80)/200 = 0.70$. Furthermore: $P_e = (200 - 100 - 120 + 2 \times 100 \times 120/200)/200 = 0.50$. The complete outcomes for the three categories are in Table 4.7.

All these outcomes are labeled as 'fair.' Agreement is highest with respect to the first category, but again the outcomes are in my view too low to proceed with the assignments by one of the raters.

The κ for two raters can also be computed based on the parts used for assignment to category of interest or not

$$\kappa = \sum_{i=1}^{c} \left(P_{o(i)} - P_{e(i)}\right) / \sum_{i=1}^{c} \left(1 - P_{e(i)}\right). \tag{4.25}$$

For the data above $\kappa_{(gh)} = (0.70 - 0.50 + 0.70 - 0.62 + 0.80 - 0.68)/(1 - 0.50 + 1 - 0.62 + 1 - 0.68) = 0.33$, which is identical to what was found before.

Cohen approach, mean assignment to category of interest, core formulas (4.26) *and* (4.27)

In the second view based on mean assignment per category (Cicchetti et al. 1978):

$$P_{o(i)} = 2f_{ii} / \left(f_{i(1)} + f_{i(2)}\right) = 2p_{ii} / \left(p_{i(1)} + p_{i(2)}\right). \tag{4.26}$$

and

$$P_{e(i)} = 2f_{i(1)}f_{i(2)} / \langle N\left(f_{i(1)} + f_{i(2)}\right)\rangle = 2p_{i(1)}p_{i(2)} / \left(p_{i(1)} + p_{i(2)}\right). \tag{4.27}$$

Here, the same holds for the weights in case one has to use them as under the Scott model.

For the first category of the data in Table 4.2: $P_o = 2 \times 80/(120 + 100) = 0.73$ and $P_e = 2 \times 120 \times 100/[200 \times (120 + 100)] = 0.5455$. The findings for the separate categories are in Table 4.8.

Table 4.8 Outcomes intracategory agreement for data from Table 4.2 based on mean assignment to category based on Cohen

Cat.	Label	P_o	P_e	κ
1	Schizophrenic	0.73	0.5455	0.40
2	Manic depressive	0.40	0.2400	0.21
3	Neurosis	0.50	0.2000	0.38

$N = 200$

Again, the outcomes are 'fair.' In the present situation, weights can be used. Here too, the kappa per category based on formulas (4.23) and (4.24) is identical to the kappa based on formulas (4.26) and (4.27); in the situation, no weights are used. The basis is:

$$\kappa_i = \left(p_{ii} - p_{i(1)}p_{i(2)}\right)/\left\{\left(p_{i(1)} + p_{i(2)}\right)/2 - p_{i(1)}p_{i(2)}\right\}. \tag{4.28}$$

4.1.4 Example with More Than Two Raters

As an example of the use of all these approaches in the situation where the number of raters is higher than 2 the data in Table 4.9 are considered. A total of 40 units are rated three times, where one of three categories could be assigned to each. With respect to the Scott view, the idea is that the raters are interchangeable, when the Cohen view is followed, there are really three different raters.

Table 4.9 40 units assigned by three different raters to one of three categories

R_1	R_2	R_3	R_1	R_2	R_3	R_1	R_2	R_3	R_1	R_2	R_3
1	1	1	2	2	2	2	2	2	1	2	1
1	1	1	2	2	2	3	3	3	1	2	2
1	1	1	2	2	2	3	3	3	1	2	3
1	1	1	2	2	2	3	3	3	1	2	3
1	1	1	2	2	2	3	3	3	2	1	2
1	1	1	2	2	2	3	3	3	2	2	1
1	1	1	2	2	2	3	3	3	2	2	3
1	1	1	2	2	2	3	3	3	3	1	1
1	1	1	2	2	2	3	3	3	3	3	2
1	1	1	2	2	2	1	1	3	3	3	2

Table 4.10 Agreement tables based on data from Table 4.9

A		R_2				B		R_3				C		R_3			
		1	2	3	#			1	2	3	#			1	2	3	#
R_1	1	11	4	0	15	R_1	1	11	1	3	15	R_2	1	11	1	1	13
	2	1	13	0	14		2	1	12	1	14		2	2	12	3	17
	3	1	0	10	11		3	1	2	8	11		3	0	2	8	10
	#	13	17	10	40		#	13	16	12	40		#	13	15	12	40

Based on these data, three agreement tables can be constructed. These tables follow in Table 4.10. These tables are used in computing the agreement in examples that follow.

The amount of observed agreement in the three tables is 0.85, 0.78, and 0.78; this makes the mean observed agreement is 0.80. In case one follows Cohen, chance expected agreement is easily found for each pair of raters based on the marginal in the tables. Next, the mean of the three outcomes is to be taken; this mean is 0.3383. If Scott was followed, the mean over all raters is necessary. For the first category, this mean is (15 + 13 + 13)/(3 × 40) = 0.3417. The means for the second and third categories are 0.3833 and 0.2750. Now the total chance expected agreement is the sum of the squares of these proportions.

4.1.5 More Than Two Raters

If there are more than two raters or ratings, the γ_3 index is based on the mean observed agreement and the mean chance expected agreement, not on the average index estimate over all pairs, which is:

$$\gamma_3 = 2\sum_{g=1}^{m-1}\sum_{h=g+1}^{m}\gamma_{3(gh)}/\langle m(m-1)\rangle. \tag{4.29}$$

The argument in favor of the observed and expected means will be explained later. The argument for not using formula (4.29) is explained by using the data from Table 4.11. Say we are looking at four pairs of raters and we know for each pair the observed and chance amount of agreement. For the moment the methods, according to which chance expected agreement is computed, are not relevant.

Based on mean observed and chance expected agreement, the index will take the value $\gamma_3 = 0.26 = (0.52 - 0.35)/(1 - 0.35)$. This is less than the outcome $\gamma_3 = 0.29$

Table 4.11 Observed and chance expected agreement for pairs of raters (hypothetical data)

	P_o	P_e	$1 - P_e$	$\gamma_{3(gh)}$
Pair				
1–2	1.0	0.5	0.5	1.00
1–3	0.5	0.4	0.6	0.17
1–4	0.4	0.3	0.7	0.14
2–3	0.5	0.4	0.6	0.17
2–4	0.4	0.3	0.7	0.14
3–4	0.3	0.2	0.8	0.13
Mean	0.52	0.35	0.65	0.29

based on the mean over all pairs. In the first view, a low or a high amount of chance expected agreement for a pair of raters is included in this ratio in the mean. In general, the difference between the outcomes will be small, unless there are outliers in the data.

Scott approach, core formulas (4.30) *and* (4.33)

We start again with the index according to Scott. The general formula is not repeated. The amount of observed agreement is found for the pair of raters g and h. Note this formula is identical to formula (4.2); only the subscript (gh) denoting the pair of raters is added:

$$P_{o(gh)} = \sum_{i=1}^{c} f_{ii(gh)}/N = \sum_{i=1}^{c} p_{ii(gh)}. \tag{4.30}$$

The proportion of observed agreement for all pairs of m raters now is:

$$P_o = \sum_{g=1}^{m} \sum_{\substack{h=1 \\ h \neq g}}^{m} P_{o(gh)}/\langle m(m-1)\rangle = 2\sum_{g=1}^{m-1} \sum_{h=g+1}^{m} P_{o(gh)}/\langle m(m-1)\rangle. \tag{4.31}$$

Following Scott, chance expected agreement is based on the use of each category over all raters. Before, we used $p_i = (p_{i(1)} + p_{i(2)})/2$ in the situation of two raters. In the situation of more than two raters, one will use

$$p_i = \sum_{g=1}^{m} p_{i(g)}/m. \tag{4.32}$$

Now one gets

$$P_e = \sum_{i=1}^{c} p_i^2. \tag{4.33}$$

As p_i refers to the use of category i over all ratings, the above formula holds not only for the comparison of assignments by two raters, but for the comparison over all ratings. It is the general formula to be used for computing chance expected agreement under the Scott model.

The above index is also referred to as κ by Fleiss, this is so because Fleiss (1971) wanted to present an extension to the comparison of more than two raters of the index by Cohen, but he did by accident so for the index by Scott, and at first, nobody was aware of this.

We had already seen that observed agreement for the data in Table 4.9 is 0.80, and chance expected agreement is $0.3417^2 + 0.3833^2 + 0.2750^2 = 0.3933$. Now the index can be computed: $\pi = (0.80 - 0.3933)/(1 - 0.3933) = 0.70$. The amount of agreement is 'substantial.'

Cohen approach, core formulas (4.30) *and* (4.34)

In the Cohen model, observed agreement is computed in the same way as in the Scott model. Chance expected agreement, however, is computed in a different way, i.e., based on the marginal distribution by each separate rater (Hubert 1977):

$$P_{e(gh)} = \sum_{i=1}^{c} f_{i(g)} f_{i(h)}/N^2 = \sum_{i=1}^{c} p_{i(g)} p_{i(h)}. \tag{4.34}$$

The chance expected agreement over all pairs:

$$P_e = \sum_{g=1}^{m} \sum_{\substack{h=1 \\ h \neq g}}^{m} P_{e(gh)}/\langle m(m-1)\rangle = 2\sum_{g=1}^{m-1} \sum_{h=g+1}^{m} P_{e(gh)}/\langle m(m-1)\rangle. \tag{4.35}$$

Above, it was already mentioned that the amount of observed agreement is again 0.80 and the amount of chance expected agreement is 0.3833. This makes $\kappa = (0.80 - 0.3833)/(1 - 0.3833) = 0.70$. Due to rounding, it looks as this is identical to the outcome found when the Scott approach was followed.

Ways of representing the data

In the situation of more than two raters, data usually are not available as a series of agreement tables, but as a data matrix. In most research situations, such a data matrix will be as the one in Fig. 4.2. The matrix has three dimensions. In the rows are the units, in the columns the raters and final dimension the categories. In the cells is the reference to the category to which the unit has been assigned by the rater. In this, matrix y_{sig} has the value 1 if unit s has been assigned by rater g to category i. In case the unit has not been assigned by the rater to the category $y_{sig} = 0$.

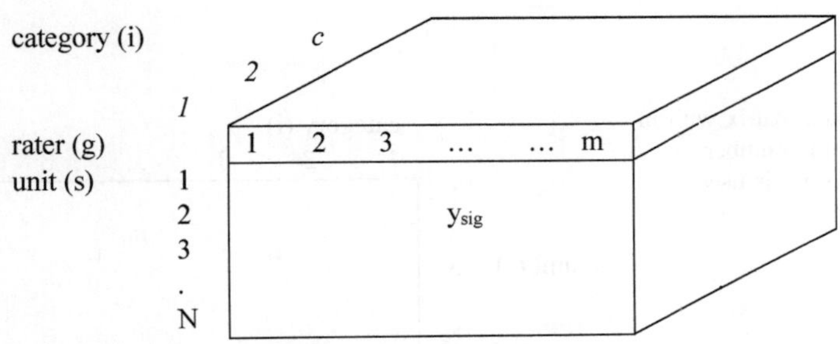

Fig. 4.2 Data matrix having dimensions for units, raters, and categories

Formulas for observed and chance expected agreement following Cohen in terms of y_{sig} have been presented by Davies and Fleiss (1982) and by Uebersax (1982). Schouten (1980, 1982a) did so for the situation in which Scott is followed.

Later in the text the fact will be used that in many situations the exact identity of the raters is not relevant, this is because all raters are considered as equal. This is in line with the view by Scott. Important however is the number of times a unit has been rated. When this situation is to be represented, one does not need information concerning the raters, but concerning the number of times the unit is rated. The matrix in which the individual raters are omitted is shown in Fig. 4.3. The situation where raters do not play a role in the computation offers new possibilities. It allows comparing assignments by raters in situations where moment of the occurrence of the target behavior cannot be predicted. This can be in a hospital situation where measurement can only be performed when a patient shows a certain behavior, say an attack of epilepsy. The target behavior, which is a specific characteristic of epilepsy, can only be measured at the moment such an attack occurs. Here, coding can only be performed by those who are present. Again, it is assumed that these raters are well equipped to perform the coding.

In this matrix, n_{si} denotes the number of times unit s has been assigned to category i. It is possible that not all units are rated the same number of times. In that case a unit is rated $n_{s.}$ times, a category is rated n_i times. In case all units are rated the same number of times, one finds $n_{s.} = m$. Complete observed agreement with respect to a unit is found when this number n_s is found in one of the cells in the row and the other cells in that row are empty. In case the number of ratings per unit can be different, so in a way if there are missing scores, n_s. can be different for each unit, at most its value is m. The least permitted value is $n_{si} = 2$; otherwise, there cannot be a comparison between ratings.

Say $f_{ij(gh)}$ is the number of units in cell $[i, j]$ of the agreement table for raters g and h, and $p_{ij(gh)}$ is the proportion of units in that cell of the agreement table for these raters. Now it is possible to indicate the relation between the data in Figs. 4.1 and 4.2. Note in the formula below not the assignments by raters 1 and 2 are considered, but the assignments by raters g and h:

$$f_{ij(gh)} = \sum_{s=1}^{N} y_{sig} y_{sjh}. \tag{4.36}$$

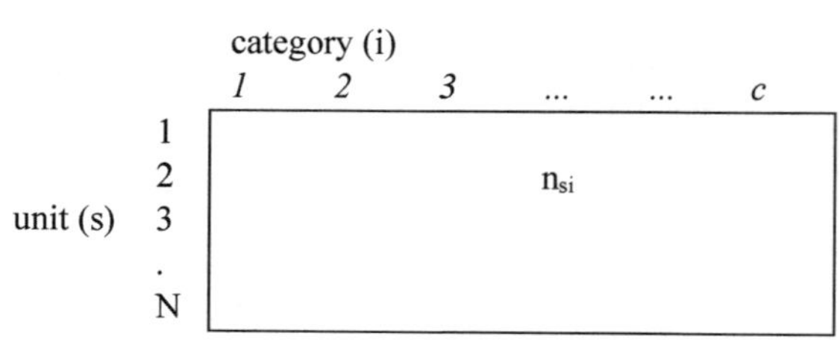

Fig. 4.3 Data matrix with in the column the number of times a category is used

The relation between the data in Figs. 4.3 and 4.2 is:

$$n_{si} = \sum_{g=1}^{m} y_{sig}, \tag{4.37}$$

and for Figs. 4.1 and 4.3

$$f_{ii} = \sum_{s=1}^{N} n_{si}(n_{si} - 1), \tag{4.38}$$

$$f_{ij} = \sum_{s-1}^{N} n_{si}n_{sj}(i \neq j). \tag{4.39}$$

The index based on mean observed and mean chance expected agreement can also be presented as

$$\gamma_3 = \sum_{g=1}^{m-1}\sum_{h=g+1}^{m} Q_{e(gh)}\gamma_{3(gh)} / \sum_{g=1}^{m-1}\sum_{h=g+1}^{m} Q_{e(gh)}, \tag{4.40}$$

here $Q_{e(gh)}$ (=1 − $P_{e(gh)}$) is the chance expected disagreement for the pair of raters g and h. The index equals the mean of the separate indices weighted by their amounts of expected disagreement. For the data in Table 4.11, we find (0.5 × 1.00 + 0.6 × 0.17 + 0.7 × 0.14 + 0.6 × 0.17 + 0.7 × 0.14 + 0.8 × 0.13)/(0.5 + 0.6 + 0.7 + 0.6 + 0.7 + 0.8) = 0.26. This outcome was found before in the discussion following Table 4.11.

Whether the Scott or the Cohen approach is followed does not matter, the above formula is used in both methods.

Scott approach, core formulas (4.43) *and* (4.46)

In case we follow the approach by Scott, the same results are found when one starts from a data matrix as in Fig. 4.1. Now only the ratings are considered and not the assignments by any specific rater. The proportion of all pairs of assignments to category i is (Schouten 1980):

$$p_{ii} = \sum_{s=1}^{N} n_{si}(n_{si} - 1)/\langle Nn_{s.}(n_{s.} - 1)\rangle, \tag{4.41}$$

All pairs assigned once to category i and once to category j ($i \neq j$) are found by using:

$$p_{ij} = \sum_{s=1}^{N} n_{si} n_{sj} / \langle N n_{s.}(n_{s.} - 1) \rangle. \tag{4.42}$$

Now the proportion of observed agreement can be computed. Note this is already the agreement over all possible pairs (so it includes both the comparison between raters g and h, but also between raters h and g). This is the reason why the formula for P_o before has been presented in two slightly different ways. Now:

$$P_o = \sum_{i=1}^{c} p_{ii}. \tag{4.43}$$

The chance expected agreement for the situation in which the identity per rater is not relevant, this is in line with the view by Scott, is based on the number of times a category is used. This number is

$$n_i = \sum_{s=1}^{N} n_{si}. \tag{4.44}$$

The proportion in which each category is used now is in the situation that all units are judged the same number of times

$$p_i = n_i / (\mathrm{Nm}) = \sum_{s=1}^{N} n_{si} / (\mathrm{Nm}). \tag{4.45}$$

The amount of chance expected agreement now is

$$P_e = \sum_{i=1}^{c} p_i^2. \tag{4.46}$$

This formula is identical to formula (4.3). Now all parts for using the general formula are known.

Conger (1980: 325) has shown that chance expected agreement according to the Scott approach is always greater than or equal to chance expected agreement based on the same data according to the Cohen approach. The difference between the two approaches is

$$\sum_{g=1}^{m} \sum_{i=1}^{c} \left(p_{i(g)} - p_i\right)^2 / \langle m(m-1) \rangle, \tag{4.47}$$

hence proportional to the squared deviation in the use of category i by the raters. Therefore, π is always lees than or equal to κ.

4.1.6 Intracategory Agreement for More Than Two Raters

Before, formulas for intracategory agreement were presented for the situation in which two raters performed the coding (formulas 4.17–4.26). When assignments by more than two raters are available, the formulas have to be extended as all pairs of raters are to be compared as computed

Scott approach, category of interest or not, core formulas (4.48) *and* (4.18)

Define $P_{o(i-gh)}$ as the observed agreement with respect to category i for raters g and h. For the first view based on assignment to the category of interest or not according to Scott, we find (Schouten 1986):

$$P_{o(i-gh)} = \left(f_{ii(gh)} + f_{\bar{i}\bar{i}(gh)}\right)/N = 1 - 2\left(f_i - f_{ii(gh)}\right)/N = 1 - 2\left(p_i - p_{ii(gh)}\right). \quad (4.48)$$

The proportion of observed agreement with respect to category i for all pairs of m raters now is:

$$P_{o(i)} = \sum_{g=1}^{m} \sum_{\substack{h=1 \\ h \neq g}}^{m} P_{o(i-gh)} / \langle m(m-1) \rangle = 2 \sum_{g=1}^{m-1} \sum_{h=g+1}^{m} P_{o(i-gh)} / \langle m(m-1) \rangle. \quad (4.49)$$

Chance expected agreement for the single category can again be formulated in terms of assigned to category i or not assigned to category i. The formula is identical to formula (4.18).

Scott approach, mean assignment to category of interest, core formulas (4.50) *and* (4.20)

For the second view based on the mean assignment to the category of interest according to Scott (Fleiss 1971):

$$P_{o(i-gh)} = f_{ii(gh)}/f_i = p_{ii(gh)}/p_i. \quad (4.50)$$

Observed agreement over all pairs of raters, so $P_{o(i)}$, is found by applying formula (4.31). Note however, this time the agreement per category is computed. The chance expected agreement for the single category i according to the Scott model is already given in (4.20).

Cohen approach, category of interest or not, core formulas (4.51) *and* (4.52)

Next follow the two views in terms of the Cohen approach. First again the situation where assignment to the category of interest or not is considered (Fleiss 1981: 220):

$$\begin{aligned} P_{\mathrm{o}(i-gh)} &= \left(f_{ii(gh)} + f_{\bar{i}\bar{i}(gh)}\right)/N = 1 - \left(f_{i(g)} + f_{i(h)} - 2f_{ii(gh)}\right)/N \\ &= 1 - p_{i(g)} - p_{i(h)} + 2p_{ii(gh)}. \end{aligned} \tag{4.51}$$

Observed agreement for category i over all pairs of raters is computed using (4.48) again. In this model, chance expected agreement on category i for raters g and h is found by using:

$$\begin{aligned} P_{\mathrm{e}(i-gh)} &= \left(f_{i(g)}f_{i(h)} + f_{\bar{i}(g)}f_{\bar{i}(h)}\right)/N^2 = \left(N - f_{i(g)}f_{i(h)} + 2f_{i(g)}f_{i(h)}\right)/N/N \\ &= 1 - p_{i(g)} - p_{i(h)} + 2p_{i(g)}p_{i(h)}. \end{aligned} \tag{4.52}$$

The proportion of chance expected agreement for all pairs of m raters on category i now is:

$$P_{\mathrm{e}(i)} = \sum_{g=1}^{m}\sum_{\substack{h=1 \\ h\neq g}}^{m} P_{\mathrm{e}(i-gh)}/\langle m(m-1)\rangle = 2\sum_{g=1}^{m-1}\sum_{h=g+1}^{m} P_{\mathrm{e}(i-gh)}/\langle m(m-1)\rangle. \tag{4.53}$$

Cohen approach, mean assignment to category of interest, core formulas (4.54) *and* (4.55)

In the second view based on mean assignment per category (Cicchetti et al. 1978):

$$P_{\mathrm{o}(i-gh)} = 2f_{ii}/\left(f_{i(g)} + f_{i(h)}\right) = 2p_{11}/\left(p_{i(g)} + p_{i(h)}\right). \tag{4.54}$$

and

Table 4.12 Outcomes intracategory agreement for data from Table 4.9

		Scott		Cohen	
	P_o	P_e	π	P_e	κ
All	0.80	0.3393	0.70	0.3383	0.70
Per category based on dichotomy					
1	0.87	0.5501	0.70	0.5496	0.70
2	0.85	0.5272	0.68	0.5262	0.68
3	0.88	0.6012	0.71	0.6008	0.71
Per category based on mean assignment					
1	0.80	0.3417	0.70	0.3409	0.70
2	0.80	0.3833	0.68	0.3821	0.68
3	0.79	0.2750	0.71	0.2742	0.71

$$P_{e(i-gh)} = 2f_{i(g)}f_{i(h)}/N\left(f_{i(g)} + f_{i(h)}\right) = 2p_{i(g)}pf_{i(h)}/\left(p_{i(g)} + p_{i(h)}\right). \quad (4.55)$$

Observed agreement and chance expected agreement over all pairs for category i are again found by using formulas (4.48) and (4.53). Here, the same holds for the weights in case one has to use them as under the Scott model.

The agreement computed for the data in Table 4.9 is presented in Table 4.12. Shown is the agreement for the total data set and per category, according to both the Scott and the Cohen views on chance expected agreement.

The index takes the value of 0.70, no matter which approach for computing chance expected agreement was followed. From the amount of chance expected agreement, it is clear however that the outcomes are not similar. The results are close to each other because there are few differences in the marginal distributions as used by the three raters. Comparable results are found for the computations per category. Using the approach based on dichotomy or on mean assignment results in different values for observed and chance expected agreement, but for the ultimate index always the same value is found. The agreement per category, no matter whether the Scott or the Cohen approach is followed, is labeled as 'moderate.' Now one can easily verify using formula (4.41) that the final π is a weighted version of all π per category.

Another example is based on a data set on injuries and fair play in sports. 87 clauses from 40 text blocks are analyzed. Raters had to indicate whether each clause contains a self-reference performing action or not. A self-reference is found in words like 'I,' 'we,' 'me,' 'our,' and so on. An example is the following sentence consisting of four clauses. The fourth clause contains a self-reference: 'He made a sliding, and as he started too late, he hit me against the ankle. I had an enormous pain.' The raters were students participating in a course, and they were not trained in any way. They had seen two examples, one of which is the one just mentioned. During the course they were just confronted with the data, the idea was to show them the difficulty of coding. The third clause contains the word 'me' (he hit me). This is not coded as no action is taken by the person.

Table 4.13 Observed agreement (P_o) for the standard and all pairs of raters G1 to H4

St	1.00									
G1	0.92	1.00								
G2	0.78	0.86	1.00							
G3	0.93	0.92	0.85	1.00						
G4	0.87	0.86	0.79	0.90	1.00					
G5	0.91	0.97	0.85	0.91	0.85	1.00				
H1	0.87	0.84	0.79	0.83	0.84	0.83	1.00			
H2	0.13	0.16	0.21	0.17	0.16	0.17	0.00	1.00		
H3	0.55	0.54	0.56	0.55	0.59	0.55	0.59	0.42	1.00	
H4	0.46	0.47	0.45	0.46	0.43	0.44	0.40	0.60	0.03	1.00
	St	G1	G2	G3	G4	G5	H1	H2	H3	H4

Table 4.14 Proportion assignments to category self-reference by each of the raters and by different groups of raters

	St	G1	G2	G3	G4	G5	H1	H2	H3	H4
$p_{1(g)}$	0.345	0.333	0.425	0.345	0.333	0.345	0.448	0.552	0.379	0.586
$p_{1(mG)}$				0.356				0.491		
$p_{1(mG+H)}$						0.416				

We will use data from 10 raters. One rater is a standard (see later). Five raters, denoted G1 to G5, belong to group G; these are students participating in one of the courses. Next follow the raters from group H; these are students participating in the course in another year. The matrix in Table 4.13 contains the proportion of observed agreement between all pairs of raters who classified the self-reference data that have been introduced above. In this matrix, St denotes the standard rater, G1 to G5 are the raters in the first group, and H1 to H4 are the raters in the second group. In total, there were 87 clauses, or following the terminology in the book, 87 units. Based on this information, the proportion of observed agreement for *almost* all empirical situations discussed can be computed.

The raters could only code whether the clause contains a self-reference or not. Below follows per rater the proportion of assignments to the category denoting that a self-reference was coded ($p_{1(g)}$). Next follow the two proportions observed agreement for the two groups of raters separately ($p_{1(mG)}$), based on the observed mean over all pairs in a group. Finally, the mean for the two groups combined ($p_{1(mG+H)}$), again based on the observed agreement over all pairs of raters, is shown. When in a specific situation the proportion assignments to self-reference are known, the proportion to no self-reference in that situation is also known, it is 1—the proportion of self-references. Based on these data, the amount of chance expected agreement in most situations that have been discussed is found easily, see Table 4.14.

Considering the five raters in group G, we find the mean observed agreement by summing the agreement between all 5 × 4/2 pairs and dividing this by the number of pairs: $P_o = 0.88$. All raters are equally qualified to do the coding task; therefore, the computation of the chance expected agreement should follow the Scott approach: $P_e = 0.356^2 \times (1 - 0.356)^2 = 0.54$. Therefore $\pi = (0.88 - 0.54)/(1 - 0.54) = 0.73$. Knowing the raters had no training at all, this probably is not a bad result. Before we had seen that the outcome 0.80 is often mentioned as a lower bound for good reliability. Using this criterion, we have to conclude however that the result found is not good enough to decide that coding was performed in a reliable way and that therefore one rater cannot be replaced by any other rater in the group. For group H, consisting of four raters, we find $\pi = (0.34 - 0.50)/(1 - 0.50) = -0.32$, an unpleasant result as the amount of observed agreement is even less than what might be expected by chance.

As the raters only used two categories, it is useless to compute the amount of agreement per category, and the outcome per category is identical to the outcome over the categories.

The data set from this example is also used in the sections that follow.

4.1.7 Elaboration Research Situations

Above the index for pairwise comparison between raters or ratings has been discussed here, the complete data set or a specific category was considered. This is what is found most in research. In some studies however, other specific research situations have been distinguished, which also need attention. These are the selection of the best or worst rater, the contribution of a new rater, the comparison of two groups of raters, and the situation where one rater is considered a standard. Also, intercategory agreement will get attention as well as the situation in which a unit can be coded more than once and the situation in which there are missing judgments.

In case one has ratings, the identity of the rater is not relevant, one only looks at the pairwise agreement for the complete data set or per category, and sometimes at the frequency in which a pair of different categories is used. Is there a combination that occurs regularly? This is all an investigator can do. In the situation where the raters themselves are important, especially when one rater is considered a standard, adaptations of the index are possible.

Selecting the best or worst rater

In the situation where more than two raters are available, it might be worth to know which rater performed best or worst. This certainly will be the case during the training phase. The question is only relevant in the situation where all raters are expected to be equal, so where the Scott approach is to be followed. The amount of chance expected agreement is not relevant now, as it is the same in each comparison. It is based on the average use of the marginal totals over all raters (Schouten 1982b). In case the view by Cohen is followed, differences between the raters might be due to the fact that different criteria played a role in the assigning process; anyway for each pair of raters, a different amount of chance expected agreement will be found. The observed agreement for all raters might be ranked. The lower the rater in the ranking, the more questionable his or her contribution to agreement. Say a is the candidate best rater and is part of the group of m raters.

$$P_{\mathrm{o}(a)} = \sum_{\substack{g=1 \\ g \neq a}}^{m} P_{o(ag)} / (m-1), \tag{4.56}$$

The only benefit of also using the amount of chance expected agreement here is that now the estimate for the index according to Scott can be computed. This allows relating the outcomes to the criterion that was posed for determining whether reliability is good or not, the order in which the raters perform will not be affected, however. The index might also be used to find the worst rater, i.e., the rater who needs further training.

One might also compute the mean estimate for the index as was already done before. Now it can be investigated whether the index above differs significantly from the mean. If so the rater who is concerned is removed from the group of raters and computations are performed again. Now at the end, a homogeneous subgroup of raters will remain. Schouten warns that one should not blindly follow the results of the significance test. One also has to consider the increase of the estimate of the index over all pairs of raters; also, practical arguments may play some role. It is also possible that a rater who is removed did not do a bad job, the rater just used other criteria.

With respect to the data in Table 4.9, we can use their presentation in Table 4.10 to compute the observed agreement for the three pairs of raters. For the pair 1 and 2, we find $P_{o(12)} = 0.85$; for pair 1 and 3 $P_{o(13)} = 0.78$; and for pair 2 and 3 also $P_{o(23)} = 0.78$. Based on these outcomes, we find for the first rater $P_{o(1)} = (0.85 + 0.78)/2 = 0.815$. In the same way $P_{o(2)} = 0.815$ and $P_{o(3)} = 0.78$. This shows the third rater performs worst.

Looking at the students in group G in Table 4.13, it turns out that observed agreement as computed by using formula (4.56) ranges from 0.90 (G1) down to 0.84 (G2); for the students in group H, however, it ranged from 0.34 (H3) down to 0.33 (H1).

Schouten (1982b: 52) presents two warnings. These concern the worst rater. The first warning is that one cannot blindly go by significance tests (these I have not presented here). The decision whether a rater belongs to the homogeneous subgroup or not should also depend on the increase in kappa and on practical considerations. The second is that the correct diagnoses are usually unknown. The results from the analysis cannot even suggest that a specific rater is making incorrect diagnoses. It is more likely that he is using different criteria than the other raters. Serious discussions about these criteria may help raters to speak the same language and to improve future interrater agreement.

Contribution by one (new) rater

Comparing the assignments by a new rater to those by the already available raters might be useful as a check to test whether there are no differences between the raters. This is especially useful when there is no standard rater. Williams (1976) refers to proficiency surveys which use reference laboratories for the evaluation of participant laboratories. He proposed an index. A condition is that the new rater has the same skills as the raters in the already known group. This implies that the marginal distribution as will be found for the new rater is supposed to be identical to the distribution found for all other raters. Therefore, chance expected agreement does not have to be considered.

Say b is the new rater; this rater does not belong to the already available group of m raters.

$$I = P_{\text{n}}/P_{\text{o}}, \tag{4.57}$$

$$P_{\text{n}} = \sum_{g=1}^{m} P_{\text{o}(gb)}/m, \tag{4.58}$$

$$P_{\text{o}} = 2\sum_{g=1}^{m-1}\sum_{h=g+1}^{m} P_{o(gh)}/\langle m(m-1)\rangle. \tag{4.59}$$

Here too chance expected agreement is not relevant when Scott is followed. It is based on the assignments by the m already known raters. In case the raters are exchangeable, the marginal totals would also apply to the new rater.

Assume we have again the data by the five students in group G who coded the self-reference. The mean observed proportion of agreement, P_o, between the 5 × 4/2 pairs is 0.88 as we had already seen before. Also assume that the rater who acted before as standard is now supposed to become the new rater. We already know that comparing this rater to the other ones results in P_n = 0.882. The ratio informing about the contribution of this new rater now is I = 0.882/0.80 = 1.002. For the other four students in group H, the mean observed agreement over the 4 × 3/2 pairs P_o = 0.34. Again, using the standard as a new rater P_n = 0.50, therefore I = 0.50/0.34 = 1.47.

In these formulas P_o is identical to the P_o's as defined before, it is the mean observed agreement between the m already known raters. P_n is the mean observed agreement between the new rater and each of the already known raters. Index I denotes the ratio between the two outcomes. If the new rater is not differing from the known ones, the outcome I = 1 should be found. The outcome $I > 1$ implies that on average the agreement between the new rater and the others is higher than the agreement among the raters in the group. In a way, the result I = 1.47 is not surprising. The new rater actually is the standard. All raters should be as close as possible to this rater. Therefore, one might expect that the agreement between this rater and each of the other raters is high compared to the comparison of any two other raters. It is expected that each rater deviates from the new rater at a specific place, but this place is different for each individual rater. Note also that this index does not say anything about the amount of agreement itself, but only about how well the new rater does compare to the already known raters.

Comparing groups of raters

Before the method to select the best rater was discussed. Schouten (1982b) has proposed a method to compare groups of raters. For this, a stepwise process was used. He started by comparing all raters together, and in each step, one rater was removed from the comparison. This was the rater who had onaverage the lowest agreement with the other raters. This is a process of backward elimination. It is not known how many times the process is to be repeated. Two answers however seem reasonable:

1. The process is to be repeated until a fixed number of raters (agreed upon in advance) has remained;
2. The process is repeated until the mean agreement over all raters is higher than some criterion value decided upon in advance.

If this second view is followed, it is less important how many raters remain. The relevant issue is that a quality criterion is met.

Instead of applying backward elimination, forward elimination might have been applied. In that case, one starts with the pair of raters having the highest score on the agreement index. Next, a third rater is looked for so that the mean agreement for the raters involved is as high as possible. In each step, a next good rater is selected. Here, one might pose additional requirements as that the outcome for the new selected rater does not differ significantly from the mean outcome for the already selected raters and shows no negative agreement with any of the already selected raters. Here too, a lower criterion for inclusion can be formulated.

This method is equivalent to the one used in cumulative scaling according to Mokken (1971) and based on Loevinger's H (Loevinger 1948), an index that compares the actual number of errors to the expected number of errors if the items in the scale would be unrelated. The method is to be preferred over backward elimination as it is possible now the control whether the requirements in the previous chapter have been fulfilled. The homogeneous subgroup of raters that results is denoted as a cluster.

In case there are a lot of raters available, it might be useful to define several clusters. This should be done in such a way that the agreement within a cluster is always higher than the agreement between clusters. For this, it is necessary to investigate how and why clusters differ.

For cluster G, consisting of m_{G} raters, an agreement index is defined which is identical to the agreement index proposed before for computing the mean agreement among more than two raters. Besides Schouten proposes an index for the agreement between clusters:

$$P_{\mathrm{o}} = \sum_{\substack{g=1 \\ g \in m_{\mathrm{G}} \\ g \notin m_{\mathrm{H}}}}^{m_{\mathrm{G}}} \sum_{\substack{h=1 \\ h \in m_{\mathrm{H}} \\ h \notin m_{\mathrm{G}}}}^{m_{\mathrm{H}}} P_{o(gh)} / (m_{\mathrm{G}} m_{\mathrm{H}}). \tag{4.60}$$

$$P_{\mathrm{e}} = \sum_{\substack{g=1 \\ g \in m_{\mathrm{G}} \\ g \notin m_{\mathrm{H}}}}^{m_{\mathrm{G}}} \sum_{\substack{h=1 \\ h \in m_{\mathrm{H}} \\ h \notin m_{\mathrm{G}}}}^{m_{\mathrm{H}}} P_{e(gh)} / (m_{\mathrm{G}} m_{\mathrm{H}}). \tag{4.61}$$

This is an inter-cluster index. All agreement indices that have been defined so far actually are intra-cluster indices.

For the students in the two groups from our example coding the self-references we had found P_{o} for the five students in group G to be 0.88; P_{o} for the four students in group H was 0.34. P_o for the $5 \times 4 = 20$ pairs from both groups $P_o = 0.50$. P_e in group G is 0.54, in group H 0.50, and between the groups 0.51.

In order to detect important sources of disagreement, one may try to divide the group of raters into several homogeneous subgroups, in such a way that the degree of interrater agreement is higher within subgroups than between subgroups. Future interrater agreement might be improved when we find out why and in which way subgroups differ in opinion.

It is possible that the coding process in an investigation is disturbed by some factor. For example, there might have been a considerable time between the coding of the first part of the data and the coding of the second part or one set of data was collected at another place that the other set might be relevant in observation studies where the setting can be relevant. The marginal probability of classification of a particular subject may depend on such facts. For example, a physician rating whether a patient has particular disease symptoms may be influenced by the overall severity of the disease. On the basis of this, one may want to assess the interrater agreement with subjects grouped into strata. Barlow et al. (1991) discussed several approaches for evaluating stratified agreement, assuming that the underlying index is common across strata but that the probability structure for each part of the data set may differ. The problem then arises of combining the strata to yield a summary 'stratified' statistic. The authors consider three weighting schemes and compare them in a simulation study: (1) equal weighting; (2) weighting by the size of the part of the data set; and (3) weighting by the inverse of the variance. In a simulation study, they compared these methods under differing probability structures and differing sample sizes for the data sets. Weighting by sample size turned out to be moderately efficient under most conditions.

Assume there are K parts of the data set, i.e., K different groups of raters; the number of units coded in group k is N_k. Weighting by sample size results in the following:

$$I = \sum_{k=1}^{K} N_k I_k / \sum_{k=1}^{K} N_k. \qquad (4.62)$$

One rater is a standard

In some situation, especially when training is going on, one may want to identify one rater as being of special interest and indicate this rater as a standard or criterion against which to compare the assignments by the other rater or raters. In this situation, accuracy is measured and not agreement. Now the γ_3 measure as either formulated by Scott or Cohen is not adequate as it is a symmetric measure. The focus has to be on the standard (Landis and Koch 1975: 157; Fleiss 1981: 233). This implies the marginal distribution as provided by the standard is essential in computing the chance expected agreement. One will get the intracategory agreement conditioned on the standard. The index below starts from this assumption. In case this view is not followed, standard and other raters get the same weight in the determination of chance expected agreement. This agreement might be based on a formula resembling (4.19) when Scott is followed or (4.23) when Cohen is followed.

Say x is the standard rater; this rater is not a member of the group of g raters who are to be compared to the standard.

$$P_{\text{o}} = \sum_{g=1}^{m} P_{\text{o}(gx)}/m, \tag{4.63}$$

Following Fleiss, it now is most appropriate to use the distribution over categories by the standard to compute the amount of chance expected agreement:

$$P_{\text{e}} = \sum_{i=1}^{c} p_{i(x)}^{2}. \tag{4.64}$$

Let us look at the example data. The assignments by the five students in group G are compared to the assignments by the standard. The observed proportions of agreement are: 0.92, 0.78, 0.93, 0.87, and 0.91 as can be seen in Table 4.13. On average based on formula (4.54), the proportion is 0.88. The marginal proportions for the standard were 0.345 for self-reference and 0.655 for no self-reference mentioned in the clause. Summing the squares gives the amount of chance expected agreement: 0.548. Therefore, the outcome of the agreement index will be $\pi = (0.88 - 0.548)/(1 - 0.548) = 0.74$. The observed proportions for the four students in group H are: 0.87, 0.13, 0.55, and 0.46, on average 0.50. The chance expected agreement is again based on the assignments by the rater. This time $\pi = (0.50 - 0.548)/(1 - 0.548) = -0.11$. The interpretation of the outcomes is as for the situation in which all pairs were compared.

Here too agreement per category can be computed. Again, the computations are conditioned on the standard.

Category of interest or not

Define $P_{\text{o}(i-gx)}$ as the observed agreement with respect to category i for rater g and standard x. For the first view based on assignment to the category of interest or not, we find:

$$P_{\text{o}(i-gx)} = \left(f_{ii(gx)} + f_{\bar{i}\bar{i}(gx)}\right)/N = 1 - 2\left(f_{i(x)} - f_{ii(gx)}\right)/N = 1 - 2\left(p_{i(x)} - p_{ii(gx)}\right). \tag{4.65}$$

The proportion of observed agreement for category i over all pairs $P_{o(i)}$ is

$$P_{o(i)} = \sum_{g=1}^{m} P_{o(i-gx)}/m. \tag{4.66}$$

Chance expected agreement for the single category can also be formulated in terms of assigned to category i or not assigned to category i:

$$P_{e(i)} = \left(f_{i(x)}^2 + f_{\neg i(x)}{}^2\right)/N^2 = \langle N - 2f_{i(x)}\left(N - f_{i(x)}\right)/N\rangle/N = 1 - 2p_{i(x)}\left(1 - p_{i(x)}\right). \tag{4.67}$$

Let us consider the assignments by the first two raters whose assignments are presented in Table 4.10. Let us assume that rater 1 is the standard. For the first category, we find $P_{o(i\_12)} = 1 - 2\ (15 - 11)/40 = 0.80$, $P_{e(i)} = (40 - 2 \times 15\ (40 - 15)/40)/N = 0.5313$. Therefore $\kappa_{1-12} = (0.80 - 0.5313)/(1 - 0.5313) = 0.57$. If we want to know the agreement between both rater 2 and 3 with standard rater 1, we first have to compute the observed agreement for rater 3 with the standard. In Table 4.10, we see that $f_{11(13)}$ is again 11. The marginal total is of course the same as before; therefore, the observed agreement must also again be as in the first comparison: $P_{o(i\_13)} = 0.80$. The mean observed agreement over the two pairs is 0.80. As the chance expected agreement does not change, we find with respect to the first category over all pairs $\kappa_1 = 0.57$.

Mean assignment to category of interest

For the second view based on the mean assignment to the category of interest:

$$P_{o(i-gx)} = f_{ii(gx)}/f_{i(x)} = p_{ii(gx)}/p_{i(x)}. \tag{4.68}$$

The proportion of observed agreement for category i over all pairs $P_{o(i)}$ is

$$P_{o(i)} = \sum_{g=1}^{m} P_{o(i-gx)}/m. \tag{4.69}$$

When the average use of the category is considered, the chance expected agreement for the single category i is

$$P_{e(i)} = p_{i(x)}^2/p_{i(x)} = p_{i(x)}. \tag{4.70}$$

Agreement per category for the example data is not computed. This is useless as there are only two categories.

Intercategory agreement

Intercategory or interclass agreement is used to investigate how often two categories (i and j) are confused. The method helps in finding systematic disagreements. Intracategory agreement, which was discussed before, informs about the category that caused most problems as agreement is lowest. This method goes further. It might be difficult for raters to distinguish between two categories, because the difference is not sufficiently clear or hard to recognize. Now one might look at how often one rater chooses category i while another rater coded category j ($i \neq j$). One will discover where problems are located by looking at the specific units involved. One might be able to improve the set of categories or the rules for assigning. This all is very useful in the pilot part of any investigation.

When comparing the assignments by more than two raters or ratings, the investigator will only find whether categories i and j are often confused, but not which rater did choose i and which one did choose j. One finds where a possible problem is located, but not why it is a problem. This is also true for the situation of two ratings (for which more than two raters were available). When only the codings by two fixed raters are compared, it will become clear how often it occurred that the one rater has chosen category i and the other rater category j in assigning one and the same unit. In case the amount of agreement is far above zero, the investigator can ask the raters to explain their choices. This will allow learning whether a specific interpretation of the issue to be scored was available. The investigator will not only learn there are differences in coding, but also why these differences exist. Note that choosing category i by one rater and category j by the other is different from choosing category j by the one and category i by the other rater.

The γ_3 index that will be found now gets the value 1 when no unit is assigned by the first rater to category i and by the second rater to category j ($i \neq j$) and will be 0 when the marginal totals of both categories i and j will be equal to the number of times the occurrence of the pair is found. The—usually asymmetric—matrix of estimates γ_3 $[i, j]$ signals due to the standardization more clearly than the raw matrix p $[i, j]$ the ordered category pairs $[i, j]$ over which confusion exists (in that situation, the γ_3 $[i, j]$ estimate is lower). The index is a special case of weighted kappa. If one wants to investigate the assignments that finally appear in cell [1, 2] of the matrix, then the weight for cell $[i, j]$ should get the value zero, while the weights for all other cells should be 1.

Due to the fact that, in general, the index is asymmetric (equality can occur incidentally), the expansion to calculating pairwise similarity between m raters on the basis of $m \times (m - 1)/2$ unordered pairs is not desirable. However, the pairwise comparison can be performed for $m \times (m - 1)$ ordered pairs of raters. Now one will only find the categories between which non-agreement (especially) occurs. It is not possible to verify whether, for example, one rater systematically scored differently from the other raters. In case the index is used to compare raters to a standard, the measure only provides an indication of the average deviation from the assignments by the rater.

In this research situation, weights cannot be used, because the researcher is interested in the (non)agreement in cell $[i, j]$. The consequence of using different weights for data from other cells causes that cells which are actually not under investigation will have a different impact on the outcomes for the cell under investigation. Let us again consider the assignments by the first two raters whose assignments are presented in Table 4.10. Each rater has used three categories.

Therefore, the agreement table has 9 cells (3 × 3), but as for intercategory agreement a category is not compared to itself there will be only 6 cells of interest, which are the 6 off-diagonal cells of the table. One will hope the frequency in each of these cells is as low as possible, by preference zero. In case this happens $p_{ij} = 0$, but P_o as is shown below will always be 1, and therefore, the estimate found for the index γ_3 will also be 1. This sounds a bit counterintuitive. The situation is found in three of the off-diagonal cells in the table. Three other cells are not empty. For these cells, we will have to compute the amount of agreement. This can only be done when we decided on the type of index to be used.

Scott approach, core formulas (4.71) *and* (4.74)

First the index in the view by Scott is presented. A high estimate for π denotes there is little disagreement with respect to categories i and j (Schouten 1982b):

$$P_{o(ij)} = 1 - \left(f_{ij} + f_{ji}\right)/N = 1 - \left(p_{ij} + p_{ji}\right), \quad (i \neq j) \tag{4.71}$$

Note that $P_{o(ij)} = P_{o(ji)}$. In case more than two raters are available

$$P_{o(ij-gh)} = 1 - \left(f_{ij(gh)} + f_{ji(gh)}\right)/N = 1 - \left(p_{ij(gh)} + p_{ji(gh)}\right), \quad (i \neq j). \tag{4.72}$$

The proportion of observed agreement over all pairs of m raters now is:

$$P_{o(ij)} = 2\sum_{g=1}^{m-1}\sum_{h=g+1}^{m} P_{o(ij-gh)}/\langle m(m-1)\rangle. \tag{4.73}$$

$$P_{e(ij)} = 1 - 2p_i p_j. \quad (i \neq j) \tag{4.74}$$

Note, the higher the estimate found for the π_{ij}-index the less disagreement between the categories i and j. This is also so for the κ_{ij}-index to be presented hereafter.

Looking at the agreement table for raters 1 and 2 as presented (Table 4.10, here rater 1 does not act as a standard anymore), we see that most differences are found with regard to categories 1 and 2. With respect to 4 + 1 = 5 units, the raters disagree, as a proportion this is 0.125. Therefore $P_{o(12)} = 0.875$. Remember that for the computation of $P_{e(12)}$ it is necessary to compute the mean assignment to each of the categories first. For category 1, this frequency is (15 + 13)/2 = 14, or as a proportion 0.35. For category 2, it is (17 + 14)/2 = 15.5, or 0.3875. Now $P_{e(12)}$ becomes $1 - 2 \times 0.35 \times 0.3875 = 0.7288$. This all makes $\pi_{(12)} = (0.875 - 0.7288)/(1 - 0.7288) = 0.54$.

With respect to the intercategory agreement on the reality claims possible and impossible as found in Table 1.1 (categories 1 and 2), we find $\pi_{12} = 0.78$. This is the worst outcome for the comparison of the four types of claims. It is believed this is due to the background of the raters, and one of them was a native American while for the other English is a second language. In the sequel of the project, the assignments by the first rater have been used.

Cohen approach, core formulas (4.75) *and* (4.76)

In the Scott approach the individual raters are not relevant, one looks at the pair of categories as used over the pairs of raters (here 3 pairs). The individual raters however are important in the approach according to Cohen, as their individual marginal distributions are used. Therefore, it is impossible to compare the assignments by more than two raters at once, and raters should be compared one by one:

$$P_{o(ij)} = 1 - f_{ij}/N = 1 - p_{ij}, \quad (i \neq j) \tag{4.75}$$

$$P_{e(ij)} = 1 - f_{i(1)}f_{i(2)}/N^2 = 1 - p_{i(1)}p_{i(2)}. \quad (i \neq j) \tag{4.76}$$

With respect to the computations, the category per rater is to be considered. Say we look at category 1 for the first rater and category 2 for the second rater. Now $P_o = 1 - 4/40 = 0.9$ and $P_e = 1 - (15/40) \times (14/40) = 0.869$. Now $\kappa = (0.9 - 0.869)/(1 - 0.869) = 0.24$. For the other comparison, we find $\kappa = (0.975 - 0.13125)/(1 - 0.13125) = 0.97$.

In case we follow the Scott approach, the conclusion can be that often it was difficult for the raters to decide whether category 1 or 2 should be used. However, we have no idea why this is so: Do the categories have too much in common, did at least one of the raters miss something (refers to the training), are the data so complex? In case the Cohen approach is followed, the results for each of the raters look much better.

The index might be seen as a special case of the weighted agreement index. Table 4.15 contains in the left part the matrix of weights for the situation four categories are used. The disagreement between categories 1 and 2 is considered.

In case two categories are easily confused, an investigator might decide to combine these categories. Schouten (1986: 455) showed for the situation of two raters that kappa will only increase if the following relation holds. The categories i and j are the one that are concerned:

Table 4.15 Weights used in computing intercategory agreement for category 1 by rater 1 and category 2 by rater 2 (assuming four categories)

		Scott approach				Cohen approach			
		Rater 2				Rater 2			
	Cat.	1	2	3	4	1	2	3	4
Rater 1	1	1	0	1	1	1	0	1	1
	2	0	1	1	1	1	1	1	1
	3	1	1	1	1	1	1	1	1
	4	1	1	1	1	1	1	1	1

$$\frac{p_{ij}+p_{ji}}{p_{i(1)}p_{j(2)}+_{j(1)}p_{i(2)}} > \frac{1-P_o}{1-P_e} = 1-\kappa, \quad (i \neq j) \tag{4.77}$$

Notice that $p_{ij} + p_{ji}$ is the observed proportion of disagreement among the combined categories, whereas $1 - P_o$ is the observed proportion of disagreement among all categories. The denominator contains the corresponding chance expected proportions of disagreement.

Warrens (2010) has shown that for any agreement table with more than two categories there exist two categories such that, when combined, the κ estimate of the collapsed agreement table, that has now $c - 1$ categories, is higher than the original κ estimate. This means it will always be possible to get a better outcome, possibly one that is above the criterion for reliability. Given the content of these categories however, it should also be realistic to combine the categories.

Multicoding

In general, units are assigned to one category. This category is part of a nominal scale and the categories should be exclusive. Nevertheless, situations are possible where units are assigned to more than one category, for example, when diagnoses are made concerning patients. In this research situation, more than two categories should be available to the rater (in the situation of two categories, one actually has the distinction 'applies' versus 'does not apply'). Now several approaches are possible. All possible combinations of available categories can be considered as new categories. This would imply that in case three categories are available 8 combinations can be made. So, there can be a lot of new categories, some of which will hardly occur, besides there will be a lot of overlap between these categories. It might also be possible that several dimensions can be distinguished within the categories after some pca or mds like technique is applied to them. Now agreement per dimension can be computed. Computations might also be performed per single category: Does the category apply or not? In this way however, one fails to consider the basic multivariate nature of the raters' choices.

Fleiss et al. (1972) proposed a model that allows determining the amount of agreement between two raters who could assign the units to several different categories. In advance, the investigator had to determine how two classifications will be weighted. Say rater g concludes a unit is to be assigned to category x or y, and next, rater h decides the assignment is to category x or z or w. Both raters mentioned category y. This fact is counted by using the weight 1. The other combinations: x–y, x–z, x–w, y–z, and y–w are all counted by using a weight in the range 0–1. The highest weight used by each rater for each category and the number of categories that have been mentioned are used as a starting point for the contribution of the unit to observed agreement and chance expected agreement.

Kraemer (1980) also considered the situation where a unit might be assigned to more than one category. The rater does not only decide whether category x or y applies, the rater also adds degrees of certainty, for example, might indicate that

Table 4.16 Table for multicoding situation

		Rater 2		
		Both	Not both	
Rater 1	Both	v_s	$v_{s(1)} - v_s$	$v_{s(1)}$.
	Not both	$v_{s(2)} - v_s$	$c - v_{s(1)} - v_{s(2)} + v_s$	$c - v_{s(1)}$
		$v_{s(2)}$	$c - v_{s(2)}$	c

category x is very certain and that category y is just possible. These indications of certainty get rank numbers. As there are c categories, c assignments are possible. The category having rank number 1 gets the weight 1, the other $c - 1$ categories receive the rank number $(c + 2)/2$, and ties are allowed. In case a unit has been assigned to two categories and there is not preference of one above the other, the two categories get the rank number 1.5 and the other categories the rank number $(c + 3)/2$. In case there is a preference, the first rank number gets the value 1 and the second one the value 2. Per unit now Spearman rank correlations can be computed for the two raters. P_o is the mean of all rank correlations, and P_e is the mean rank correlation over all possible pairs of units. The method can be extended to m raters.

According to Kupper and Hafner (1989), the choice of an appropriate agreement statistic depends on a rater's a priori knowledge regarding the prevalences of the attributes in the population under study. They argue that if such knowledge is available a kappa-type statistic might be appropriate, but if there is no such knowledge, their model for chance expected agreement should be used.

Say a unit can be scored in each category. In case this happens there will be c assignments, so if there are six different categories in theory six assignments are possible. Per category one can look at whether this category was selected or not. For unit s one gets Table 4.16 given the assignments by two raters. In this table v_s is the number of identical categories used by both raters, $v_{s(1)}$ is the number of categories used by the first rater and $v_{s(2)}$ is the number of categories used by the second rater.

Kupper and Hafner now take the position that the amount of observed agreement for unit s and category i can be defined in two ways: (1) one in which explicit credit is given to the categories that are not used by either rater and (2) one in which only categories are considered that are used by at least one of the raters. Following the first way observed, agreement is defined as:

$$P_{o(s)} = v_s/\max\left(v_{s(1)}, v_{s(2)}\right). \tag{4.78}$$

Following the second way, looking at the table's diagonal, it is defined as

$$P_{o(s)} = \frac{v_s + c - v_{s(1)} - v_{s(2)} + v_s}{c} = \frac{c + 2v_s - v_{s(1)} - v_{s(2)}}{c}. \tag{4.79}$$

This latter formula is most in line with the indices that have been discussed so far. The overall observed agreement between the raters is

$$P_{\mathrm{o}} = \sum_{s=1}^{N} P_{o(s)}/N. \tag{4.80}$$

If both raters make their assignments for the *s*th unit completely at random, the null distribution of v_{s} is central hypergeometric. Now a conditional model for chance expected agreement on the *s*th unit can be formulated, which derives from an underlying unconditional model that assumes $v_{\mathrm{s}(1)}$ and $v_{\mathrm{s}(2)}$ are independent. Kupper and Hafner elaborate their method however on the other way. The expected agreement for unit *s* is

$$P_{\mathrm{e}(s)} = \min\left(v_{\mathrm{s}(1)}, v_{\mathrm{s}(2)}\right)/c. \tag{4.81}$$

The expected agreement between the raters over all categories now becomes

$$P_{\mathrm{e}} = \sum_{s=1}^{N} P_{\mathrm{e}(s)}/N. \tag{4.82}$$

This all results in an index *C* for comparing the assignments by two raters:

$$C = (P_{\mathrm{o}} - P_{\mathrm{e}})/(1 - P_{\mathrm{e}}). \tag{4.83}$$

An interesting special case of this formula results when each rater must select only one category to describe the *s*th unit. For this situation $v_{\mathrm{s}(1)} = v_{\mathrm{s}(2)} = 1$ for all *s*. Now the index simplifies to

$$C = (cP_{\mathrm{o}} - 1)/(c - 1). \tag{4.84}$$

This is the index by Holley and Guilford (1964).

Assume two raters can use six different categories. With respect to unit s, the first rater assigns the categories 1, 2, and 3 to the unit. The other rater uses the categories 1, 2, 4, and 5. Both raters used categories 1 and 2, therefore $v_{\mathrm{s}} = 2$. The first rater used three categories, so $v_{\mathrm{s}(1)} = 3$. Following the same reasoning $v_{\mathrm{s}(2)} = 4$. Now we can, using formula (4.78), compute $P_{\mathrm{o}(s)} = 2/\max(3, 4) = 2/4 = 0.5$. Using formula (4.81), we find for this unit $P_{\mathrm{e}(s)} = \min(3, 4)/6 = 3/6 = 0.5$.

In case there are more than two raters, the index can easily be extended in a way similar to the formulas in (4.31) and (4.35). For more details, the reader is referred to the original articles.

Missing judgments

It is possible that all units have not been rated the same number of times. This can be so because at the moment at which the behavior of interest occurred not all desired raters were available or because a rater failed to do his job. In case the investigator is only interested in the ratings, the problem actually already is solved. In the indices over ratings, the total number of ratings per unit is given by $n_{s.}$. A unit must be rated at least twice, so $2 \leq n_{s.} \leq m$.

If the identity of a rater is relevant however one can compute the agreement among all pairs and see that comparisons are based on a different number of units. In case pairwise agreement is to be computed, only those units will be included in the analysis for which scores by all raters are available.

When raters in case of doubt do not assign a category to a unit will be that the index that is finally computed will be artificially high compared to the results one would get when all raters have been rated.

Schouten (1982b) proposed to solve this problem by using specific weights. Given c categories have been used, he proposes to introduce a category $d = c + 1$ for the missing judgments. A matrix of weights should be used now in which:

$$w_{id} = w_{di} = w_{dd} = 0. \tag{4.85}$$

The weighted version of the index is computed. However, two problems might be connected with this way of weighting: (1) The weighted agreement coefficients will probably get lower estimates. There is no mathematical proof for this, the assertion is based on results from a case study (Schouten 1982b: 54); (2) category d does not contain information if the missing scores can be considered as an a-select sample from the $m \times N$ units or from the N units for each rater.

Following Scott

$$p_i = n_i/(Nn_{s.}) = \sum_{s=1}^{N} n_{si}/(Nn_{s.}), \tag{4.86}$$

where $n_{s.}$ is the number of times unit s is judged. Each unit should be judged at least twice. This formulation can be used in any formula based on the Scott approach.

In the situation where an investigator may want to compare estimates found for the kappa coefficient, it is possible that some coefficients are biased upward as one or more raters have not classified all units. The units however belong to one of the available categories. There is especially a problem if the kappa estimates are not biased to the same extent. For this situation, Schouten (1986: 464) proposes the following solution.

Suppose M of the N units have been classified by both of two particular raters. For these M units, P_o denotes the observed and P_e denotes the chance expected proportion of agreement. Suppose further that, if the remaining N–M units also had been classified by both raters, for these N–M units the observed proportion of agreement would have been equal to the chance expected proportion of agreement P_e. This leads to an adjusted observed proportion of agreement

$$P_{o\_adj} = \langle P_o + (N - M)P_e \rangle / N. \tag{4.87}$$

and an adjusted kappa estimate

$$\kappa_{\text{adj}} = \left(P_{\text{o\_adj}} - P_{\text{e}}\right)/(1 - P_{\text{e}}) = M(P_{\text{o}} - P_{\text{e}})/N(1 - P_{\text{e}}) = M\kappa/N. \tag{4.88}$$

Since the assumption above may be somewhat pessimistic, this adjusted kappa estimate must be considered a lower bound.

4.1.8 Simultaneous and Majority Agreement

Before it was already indicated that similarity, all raters assign a unit to the same category, and majority, at least k out of the m raters assign a unit to the same category, agreement is not advised in the situation in which more than two raters are used. One should use pairwise agreement. For completeness however, it is necessary to present the formulas.

With respect to simultaneous agreement, one would not start from a two-dimensional table (rater by rater), but from a table having as it were as much dimensions as there are raters. In this situation, observed agreement can be defined as the proportion of units on the diagonal in the table, similar to formulas (4.2) and (4.6) before. The proportion expected agreement is based on the marginal totals, similar to formulas (4.3) or (4.7).

$$P_{\text{o}} = \sum_{i=1}^{c} f_{i_1 i_2 \ldots i_m}/N = \sum_{i=1}^{c} p_{i_1 i_2 \ldots i_m}, \tag{4.89}$$

where $f_{i_1 i_2 \ldots i_m}$ denotes the number and $p_{i_1 i_2 \ldots i_m}$ the proportion of agreeing assignments by all m raters to category i. Read $f_{i(g)}$ again as the number and $p_{i(g)}$ as the proportion of assignments by rater g to category i, then the average number resp. proportion of assignments to category i is

$$f_i = \sum_{g=1}^{m} f_{i(g)}/m; \quad p_i = \sum_{g=1}^{m} p_{i(g)}/m. \tag{4.90}$$

Now expected agreement according to the reasoning by Scott is

$$P_{\text{e}} = \sum_{i=1}^{c} (f_i/N)^m = \sum_{i=1}^{c} p_i^m. \tag{4.91}$$

and according to the reasoning by Cohen

$$P_{\mathrm{e}} = \sum_{i=1}^{c} \prod_{g=1}^{m} (f_{i_g}/N) = \sum_{i=1}^{c} \prod_{g=1}^{m} p_{i_g}. \tag{4.92}$$

These amounts of observed and expected are to be entered into the general formula (4.1). The computation of the amount of agreement might also be based on a data matrix as presented in Fig. 4.3. Here n_{si} is relevant, the number of assignments by rater s to category i. Define

$$a_s = \sum_{i=1}^{c} a_{si}, \tag{4.93}$$

where $a_{si} = 1$ in case $n_{si} = \mathrm{m}$, otherwise $a_{si} = 0$. Now

$$P_{so} = \sum_{s=1}^{N} a_{si}/N. \tag{4.94}$$

is the proportion observed agreement within the judgments. The proportion of expected agreement is defined as

$$P_{\mathrm{e}} = \sum_{i=1}^{c} P_i^m. \tag{4.95}$$

The results found in (4.7) and (4.8) have to be substituted in the formula for kappa. Ross (1977) and Hubert (1977) defined the index following Cohens view. Many authors paid attention to the index. Ross also presented a formula in which weights are used.

In majority agreement, there is already a contribution to agreement in case at least k out of the m judgments of a unit of analysis are to one and the same category. The formulas are again based on a data matrix as in Fig. 4.3. For the computation of the amount of observed agreement, formulas (4.6) and (4.7) are used again, with the difference that now $a_{si} = 1$ in case $n_{si} \geq k$, and otherwise again $a_{si} = 0$. A condition with regard to k is that $k \geq (m + c - 1)/\mathrm{c}$, and $k > m/2$; otherwise, a_{si} will by definition take the value 1, which means that one will always find that there is perfect agreement among the judgments. As an example consider the situation in which $k = 2$, $m = 5$, and $c = 4$. Because there are five judgments and four categories, at least two units have to be assigned to the same category. This satisfies the requirement $k = 2$, so there is agreement with regard to the unit of analysis. This, however, holds for any unit of analysis. Therefore, the condition is necessary.

The amount of expected agreement following Scott is

$$P_{\mathrm{e}} = \frac{1}{c} \sum_{r=k1}^{m} \sum_{i=1}^{c} \binom{m}{r} p_i^r (1 - p_i)^{m-r}. \tag{4.96}$$

The values for P_o and P_e should again be substituted in formula (4.1).

Conger (1980) has presented a formula for the computation of agreement in case just k out of the m judgments per unit of analysis are into the same category, and Craig (1981) presented a formula for the two out of three codings situation.

The amount of expected agreement P_e according to the view by Cohen will only be explained in terms of three raters in the situation that at least two out of the three codes are in the same category

$$P_e = \sum_{i=1}^{c} p_{i..}p_{.i.}p_{..i} + p_{i..}p_{.i.}(1 - p_{..i}) + p_{i..}(1 - p_{.i.})p_{..i} + (1 - p_{i..})p_{.i.}p_{..i}. \quad (4.97)$$

4.1.9 Supposed Problems Due to Chance Correction

The γ_3-index as a measure of agreement between raters is widely used, and it has some limitations and disadvantages, however. These will be discussed here, and possible solutions will be mentioned.

Chance expected agreement is defined as agreement to be expected based on the marginal distributions. This can be the distribution over all raters or the distribution for each of the separate raters as was discussed before. The formulation assumes that the raters made a complete guess. Based on this assumption, the investigator estimates the amount of agreements. In the final index, a correction is made for this amount. This estimate is based on complete statistical independence of the raters. It means that the raters guessed completely with regard to every unit, but also that they guessed with probabilities that match the marginal proportions of the observed ratings.

The question is whether this assumption is realistic. In reality, there will be doubts with regard to the assignment of a number of units. This implies guessing is partial. Therefore, the correction made is not a correct one. The basic logic behind the view on chance-corrected agreement is flawed. The problem is presented for the situation in which two raters have assigned the units to only one of two categories. Cohen was aware of the problem, but did not take it too seriously: 'each judge in fact distributes his judgments over the [c] categories, he does not simply lump them in one' (Cohen 1960: 40).

The first problem has to do with the relative probability of occurrence of each of the categories, i.e., following Scott f_1 is much larger or smaller than f_2, this is the

Table 4.17 100 units assigned by two raters (hypothetical data)

		Rater 2 (standard)		
	Applies	1	0	#
Rater 1	1	4	16	20
	0	16	64	80
	#	20	80	100

problem of 'prevalence.' The second problem is that the frequency of occurrence of the two categories is different according to the two raters, i.e., $f_{i(1)} \neq f_{i(2)}$, this is usually indicated as 'bias.' It should be clear that this problem only plays a role when Cohen's line of reasoning is followed with respect to interchangeability of raters. In Scott's model, a comparison of the assignments per rater is not the case.

One paradox in the outcome of the agreement index, described among others by Feinstein and Cicchetti (1990), and due to prevalence, tells that unbalanced marginal totals produce higher estimates of the index than more balanced totals. Zwick (1988) even mentions that this can happen when the raters have identical marginal distributions. An example where this occurs is presented in Table 4.17, where 100 units have been assigned by two raters, code *1* is used to indicate that the attribute that is investigated applies and code *0* denotes that the attribute does not apply.

For the data in this table, one would find as value of the γ_3-index zero, no matter how chance expected agreement is defined, while the proportion of observed agreement is 0.68.

The other paradox in the outcome of the agreement index, described by the same authors and due to bias, states that the proportion of observed agreement can be very high, but nevertheless one finds a relatively low estimate for the outcome of the index. This is due to the value found for the amount of chance expected agreement. Shrout et al. (1987) argue that 'a major strength of κ is precisely that it does weigh disagreements more when the base rate approaches 0% or 100%,' whereas Grove et al. (1981) contend that κ penalizes too harshly for disagreements when base rates are either very low or very high. This holds for all γ_3-type indices.

Let us look at a hypothetical example in order to get some feeling for the effect of prevalence. Assume two raters who classified units on the dichotomy 'something applies' or 'something does not apply.' They do this for several items. In the existing literature, only this situation is considered. Each time the proportion of observed agreement found is 0.9. With respect to the distribution of assignments over the categories, it turns out that for the first item both raters each assigned 0.5 of the assignments to the apply-category, for the second item this proportion is 0.4, for the third 0.3, and so on. The results found for the agreement index are in Table 4.18. The amount of chance expected agreement found for the coding of item 5 is $0.1 \times 0.1 + 0.9 \times 0.9 = 0.82$.

Table 4.18 Example of the effects of a skewed distribution of the marginal (hypothetical data)

Item	P_o	Proportion apply	Proportion not apply	P_e	γ_3
1	0.9	0.5	0.5	0.50	0.80
2	0.9	0.4	0.6	0.52	0.79
3	0.9	0.3	0.7	0.58	0.76
4	0.9	0.2	0.8	0.68	0.69
5	0.9	0.1	0.9	0.82	0.44

The more skewed the distribution of the marginal totals, the lower the estimate the chance-corrected index will take. For item 5, one finds the outcome 0.44, while the proportion of observed agreement is 0.90, which is usually seen as good.

There are three solutions to overcome the problem:

1. Make that skewed distributions are not possible;
2. Use an index in which distributions do not play a role or at least not such an important role;
3. This is part of the game. One category hardly occurs and when it occurs and raters disagree then this is a real problem and has to be punished severely.

The first solution mentioned depends on the topic or the design of the study. Often it is not realistic. In many studies, extreme situations are investigated. These are situations that do not occur that much, so prevalence is to be expected. Before, we already referred to Spanjer et al. (2009), who investigated types of physical activities that are usually unproblematic. The problematic cases hardly occur. The data in Table 1.1 were about types of reality claims as found in texts. In case the question would have been whether sentences in texts contain such claims or not, the distribution would have been really skewed, just because these claims do not occur that often. It would be very nice in case categories would occur in more or less the same frequency, but it is impossible to demand such.

The second solution asks for new indices.

Investigators have looked for indices to be used in the situation of skewed distributions. Hereafter, we will see that these investigators went back to indices of type γ_2 or worse. For this reason, this is not the perfect solution. Here, one might also look at explicit models of rater decision making (Uebersax 1987). This would require both a theoretically defensible model and sufficient data to empirically verify conformity of observed data with the model. Now one would have to m*odel* rater agreement (Agresti 1992; Uebersax 1993), rather than merely calculating a simple index. Such models will be discussed shortly in a later section. Indices as intended will be discussed in the subsections hereafter.

The third solution, accepting a low outcome, is in my view most realistic, although it is clear that it does not make an investigator very happy. Assume the situation where the category of interest is used only a limited number of times as it hardly occurs. Also assume that if it seems to occur the raters disagree. Adapting indices now is fooling oneself. The already available indices make very clear that the present situation is not good. The investigator should try to find out why the assignments are to different categories. A possibility is simply clicking the wrong button, but it might also be that it is very difficult to recognize whether all requirements for a specific coding were at present or it might simply be difficult to recognize the target behavior. The low estimate for the index emphasized the difficulty of assigning correctly the unit that occurs only few times.

Prevalence

Several indices have been proposed to be used in case the distributions are very skewed. The indices are based on prevalence and bias. However, the indices are not that new. Most of them are already known under some other name and are used as similarity coefficient in cluster analysis.

When a rater is compared to a standard, especially when this rater is a diagnostic test, one is investigating accuracy. Many investigators estimate now the *sensitivity* and the *specificity*. Sensitivity (also called recall rate in some fields) measures the proportion of actual positives which are correctly identified as such (e.g., the proportion of sick people who are correctly identified as having the condition). Specificity measures the proportion of negatives which are correctly identified (e.g., the proportion of healthy people who are correctly identified as not having the condition. With these measures, no adjustment is made for possible chance expected agreements between the rater and the standard. Assuming rater 2 in Table 4.17 is the standard, one will find specificity = $f_{11}/f_{1(2)}$ = 4/20 = 0.2 and sensitivity = $f_{00}/f_{0(2)}$ = 64/80 = 0.8.

A similar reasoning might be used for assessing agreement between two raters or tests. One can look at positive agreement (PA), the choice for category 1 in the table above, or negative agreement (NA), the choice for category 0. These are calculated as follows:

$$\mathrm{PA} = 2f_{11}/\left(f_{1(1)} + f_{1(2)}\right), \tag{4.98}$$

$$\mathrm{NA} = 2f_{00}/\left(f_{0(1)} + f_{0(2)}\right). \tag{4.99}$$

Positive agreement, for example, estimates the conditional probability, given that one of the raters, randomly selected, makes a positive rating, the other rater will also do so. For the data in Table 4.17, we find PA = 2 × 4/(20 + 20) = 8/40 = 0.2 and NA = 2 × 64/(80 + 80) = 128/160 = 0.8. Both indices are closely analogous to specificity and sensitivity. By verifying that both PA and NA are acceptably high, one is protected against unfairly capitalizing on extreme base rates when evaluating the level of rater agreement. Both indices had already been proposed in another context by Dice (1945).

The denominator of this prevalence formula consists of all judgments into the positive category; therefore, the judgments that are positive according to both raters are counted twice. This is considered in the nominator, and here, they are also counted twice. By doing so however, the final outcome is overestimated. More realistic is here the coefficient of Jaccard (Anderberg 1973). Here, the denominator consists of the judgments that are positive according to at least one of the raters, and the nominator contains the judgments on which both agree:

$$\mathrm{Jaccard} = f_{11}/(f_{11} + f_{10} + f_{01}). \tag{4.100}$$

Assuming the table that results after the coding task has been performed shows that each cell contains 10 judgments. The values that will be computed are: PA = (10 + 10)/(2 × 10 + 10 + 10) = 0.5 and Jaccard = 10/(10 + 10 + 10) = 0.33. This last result remains closer to the data available. With respect to both situations, the judgments in cell f_{00} of the table are not considered. This concerns the frequency of the judgments and also the reason why they are there.

Byrt et al. (1993) suggest that the best estimate from the sample for the probability of category 1 for the whole population would be the mean of the marginal totals $f_{1(1)}$ and $f_{1(2)}$ would be their mean. Similarly, the best estimate of the probability of the other category is the mean of $f_{0(1)}$ and $f_{0(2)}$. Byrt and colleagues label the difference between the two probabilities the Prevalence Index (PI). It is estimated by

$$\mathrm{PI} = (f_{11} - f_{00})/N. \tag{4.101}$$

This index takes values from −1, when $f_{11} = 0$ and $f_{00} = N$, to +1, when $f_{11} = N$ and $f_{00} = 0$. It is equal to 0 when both categories are equally probable. For the data in Table 4.17, we find PI = (4–64)/100 = − 0.6.

Bias

Bias occurs when the two raters do not use the two categories with the same frequency; i.e., there is no homogeneous use of categories. In order to get grip on the difference, Byrt et al. (1993) define the Bias Index (BI) to be equal to the difference in proportions of units in the preferred category (here category 1) for the two raters and estimate it by

$$\mathrm{BI} = (f_{11} + f_{10})/N - (f_{11} + f_{01})/N = (f_{10} - f_{01})/N. \tag{4.102}$$

Actually, the authors should also have included in the formula which rater comes first. In the formula above a specific rater comes first, would the other rater have come first the formula would have been $(f_{01} - f_{10})/N$.

The absolute value of BI has a minimum of 0 when $f_{10} = f_{01}$ and a maximum of 1 when either $f_{10} = N$ or $f_{01} = N$. The Bias Index is equal to zero if and only if the marginal proportions are equal. In this case $f_{i(1)} = f_{i(2)}$. This is found in our data in Table 4.17: BI = (16 − 16)/100 = 0.

According to some these indices are clear (Uebersax 1987), others still want to continue with a correction for chance expected agreement, but now this has to include prevalence and bias.

Correcting for prevalence and bias

Based on the above, some authors claim to have found a solution for the problem by developing new indices. The first index is bias-adjusted kappa (BAK)) (Byrt et al. 1993: 424). It gives the value of kappa that results if cells f_{10} and f_{01} in the table are both replaced by their average, $m = (f_{10} + f_{01})/2$. BAK is in fact Scott's π, although it is derived from a different point of view. An index that goes further is prevalence-adjusted bias-adjusted kappa (PABAK) (Byrt et al. 1993: 425). In this index, the values in cells f_{11} and f_{00} have been replaced by their means $[(f_{11} + f_{00})/2]$, as well as the values in cells f_{10} and f_{01}. The table is symmetric, the total proportion for each category is 0.5, and therefore, the amount of chance expected agreement also is 0.5. Actually, on the diagonal we still find the sum of the units in cells f_{11} and f_{00}. Therefore, the PABAK index is defined as

$$\text{PABAK} = \langle(f_{11} + f_{00})/N - \frac{1}{2}\rangle / \left(1 - \frac{1}{2}\right) = 2P_o - 1. \tag{4.103}$$

Again, for the data in Table 4.17: PABAK = 2 × (4 + 64)/100 − 1 = 0.36. The index, however, is not that new. It is an index of the type γ_2 that is discussed before, and it is identical to the G-index Holley and Guilford (1964) and to the Random Error coefficient (Maxwell 1977), but also to the earlier discussed coefficient by Bennett et al. (1954). As PABAK is used for dichotomous data, it is also equal to the Rand coefficient (the proportion on the diagonal: $[f_{11} + f_{00}]/N$) (Anderberg 1973) minus 0.5.

The relation to the γ_3 index is as follows:

$$\gamma_3 = \frac{\text{PABAK} - \text{PI}^2 + \text{BI}^2}{1 - \text{PI}^2 + \text{BI}^2}. \tag{4.104}$$

It can be seen that, unless PABAK = 1, the larger the absolute value of BI, the larger is γ_3 (for PI constant), and the larger the absolute value of PI, the smaller is γ_3 (for BI constant). If both bias and prevalence effects are present, then the result may be that γ_3 is larger or smaller than PABAK, depending on the relative size of BI and PI.

These indices BI, PI, and PABAK should not be considered as replacements for especially the κ-index. The κ-index is as it were decomposed into components reflecting observed agreement, bias, and prevalence (Byrt et al. 1993: 423). These components should be discussed when comparisons are made between agreement studies.

High outcomes are always wonderful, and investigators always hope to find such results. Chance correction might be responsible for not so beautiful outcomes. But is this a good argument for stepping back from a γ_3-index type to a γ_2-index? For me, the situation of disagreement would be very alarming, especially when one looks at an attribute that does not occur often. Here, it seems very undesirable to me that there is also no agreement among the reviewers. The occurrence of disagreement is severely punished in the indices of type γ_3. In my view, disagreement in this situation of strong prevalence is undesired and actually invites the investigator to act. This is especially true in the medical profession where many decisions relate to questions about life and death. Undesired results are as it were puffed away. The demands that are being made are decreased. This should not happen.

Investigators should use indices of type γ_3. These show the most realistic outcomes. Indices developed for situations in which prevalence or bias play some role do not consider the seriousness of the disagreement and should not be used.

4.1.10 Test on Homogeneous Use of Categories

An investigator might want to test whether the marginal distributions are different or not. One of the assumptions of the agreement index according to Scott is that these marginal distributions do not differ. The purpose of training the raters is to realize that they all perform the coding in the same way, which should have as a result that the marginal distributions are equal for each rater.

A test to investigate whether there is a difference between the marginal totals by two raters has been proposed by Maxwell (1970). In another context, the index was earlier already proposed by Stuart (1955); it is based on the McNemar's test, a statistical test used on paired nominal data to determine whether the row and column marginal frequencies in a table are equal. The test starts with a vector $\underline{d}$, in which the ith element presents the difference between the numbers of units that both raters have assigned to category i, $d_i = f_{i(1)} - f_{i(2)}$. Assuming a multinomial distribution of the d's, it should be true that $E(d_i) = N[\rho_{i(1)} - \rho_{i(2)}]$. Usually, one tests whether the marginal distributions are different, in that case the null hypothesis to be tested is $E(d_i) = 0$. As all d_i sum to zero, only the first $c - 1$ elements in the vector are relevant; this is because the cth element is fixed. The estimated covariance matrix for the vector $\underline{d}$ is $\underline{S}$. Now the following expression can be used to test whether the two distributions differ:

$$\underline{d}'\underline{S}^{-1}\underline{d}. \tag{4.105}$$

This expression is asymptotically χ^2 distributed with $c - 1$ degrees of freedom. For the data in Table 1.1, we find $\chi^2 = 26.40$; $df = 4$; $p = 0.00$. This means that the differences d_i are not that small that they can be accounted for in terms of random error. The marginal distributions of both raters differ. According to Zwick (1988: 377), 'if one rejects the hypothesis of marginal homogeneity, one need go no further. One can express the degree of disagreements between raters in terms of the discrepancies between their marginal distributions.'

Table 4.19 Outcomes test on homogeneity marginal distributions according to Fleiss and Everitt (1971)

Cat.	Estimate	Confidence interval			
		Scheffé		Bonferroni	
		Lower	Upper	Lower	Upper
1 possible	0.07	0.03	0.10	0.05	0.09
2 impossible	−0.05	−0.08	−0.02	−0.07	−0.03
3 inevitable	0.00	−0.01	0.01	−0.01	0.01
4 contingent	0.00	−0.01	0.01	−0.01	0.00
5 not coded	−0.01	−0.04	0.01	−0.03	0.00

$N = 487$, $p = 0.0$

Fleiss and Everitt (1971) have proposed another method where a confidence interval is computed around the separate d-values based on earlier work by Scheffé and by Bonferroni. The desired outcome $d_i = 0$ for the marginal distribution from the data in Table 1.1 is only in the interval for the last three categories as is shown in Table 4.19, not in the ones for the first two categories, the ones that contain most units.

The test by Maxwell shows the two marginal distributions are different; the test by Fleiss and Everitt shows that the first rater used the category ‘possible’ more than the second rater did, and that the second rater used the category ‘impossible’ more than the first rater did. The other three categories have been used with the same frequency. These results show that the distributions used by the two raters are not identical. In a way the not coded category is questionable, it measures something else than the other categories (here by definition agreement is impossible). Leaving the category out has no effect on the results found for the other categories. The value which the χ^2-index takes increases very much.

One has to be careful in interpreting these results anyway. For the data in Table 4.2, we find a far from significant χ^2 estimate and all confidence intervals according to both Scheffé and Bonferroni range from a value below zero to one above zero. However, 20 out of the 100 observation (20%) cannot be assigned by the raters into the same category. For the data above, this is only true for about 17% of the data, and this is a little bit less. The deviations however are found in relatively few categories. In the further analyses in which these data have been used, the investigators used the coding by the rater who was a native speaker. His understanding of the use of the claims possible and impossible is least questionable.

Landis and Koch (1977) proposed a model of interrater agreement based on a general approach for the analysis of multivariate categorical data, and this test does not start from the assumption that the null hypothesis is true.

4.2 A Priori Method of Coding

If one is willing to compute the agreement in assignments, one research situation is of great importance, and will be mentioned here, separate from possible other empirical research situations. It has to do with whether or not the response categories are known when the raters start their assigning task. If the categories are known, they are equal for all raters, and all units would be assigned to one of these categories. A different situation occurs when the response categories have to be developed by the raters during the assigning process. In this situation, each rater may finish with a different number and set of categories. This situation arises in pilot studies, where the investigator wishes to find a set of response categories that will be used in the main investigation.

In the first situation, it is said that the a posteriori method of coding is applied: Assigning starts when the set of categories has already been determined. Up to here, this situation has been considered. In the other situation, the a priori method of coding is applied: Assigning starts even before the sets of categories have been determined (Montgomery and Crittenden 1977). Both methods demand that agreement is determined in another way, because the categories may differ if the a priori method of coding is used.

Take two raters who have each assigned a sample of N units to the categories of a (nominal) scale. The raters may have used different scales, each also consisting of a different number of categories. The categories used by rater *g* are denoted by g_i, and those by rater *h* by h_j. Say f_{ij} denotes the number of units assigned by rater *g* to category g_i, and by rater *h* to category h_j. Categories having the same sequence number do not necessarily have to have the same content, e.g., g_1 does not have to refer to the same content as h_1. Rater *g* has used *r* categories, and rater *h* worked with c categories; *r* and *c* do not have to be equal either.

The information in the table containing all assignments can be analyzed by means of a comparison of the classifications of all pairs of units. For each rater, it holds that the two units of a pair are classified either into the same category, or into different ones. To give an example, suppose that four units have been assigned as follows:

unit *w* g_1 h_2
unit *x* g_2 h_1
unit *y* g_2 h_1
unit *z* g_3 h_1

Both raters take the position that the units *x* and *y* are assigned to the same category as separately developed. Concerning this pair, there is complete agreement. Concerning the pair *x* and *z,* this is not true. According to rater *h,* the units must be assigned to the same category, but according to rater *g* this is not true. There is no agreement here. Finally, it is possible that both raters take the position that the observations in a pair have to be assigned to different categories. In this case, there is again agreement between the raters. This is true for the pair consisting of the units *w* and *x*. Given *N* units, there are N (N − 1)/2 pairs of units: These can be placed in a table as given in Table 4.20.

In cell A, agreement means agreement in classifying into the same categories, while agreement in cell D means agreement in classifying into different categories. For the data in Table 4.20, the meaning of the original categories is of no importance; this meaning may be different for the two raters, and it is not the verbal labeling but only the assignment result that determines the outcome.

Table 4.20 Table resulting from comparison of $N (N - 1)/2$ pairs of units

Rater g	Rater h	
	Same category	Different category
Same category	A Agreement (same)	B Non-agreement
Different category	C Non-agreement	D Agreement (different)

Popping (1983a) has proposed the D_2-index to be used in this situation. The index starts from cell A in Table 4.20. Here is agreement with respect to the pairs that are compared. Cell D is not used. The cell shows agreement on a different assignment, but gives no information with regard to the difference:

$$D_2 = (D_o - D_e)/(D_\mathrm{m} - D_e), \tag{4.106}$$

where

$$D_o = \sum_{i=1}^{r} \sum_{j=1}^{c} f_{ij}(f_{ij} - 1)/\langle N(N-1)\rangle. \tag{4.107}$$

$$D_r = \sum_{i=1}^{r} f_{i(1)}(f_{i(1)} - 1)/\langle N(N-1)\rangle, \tag{4.108}$$

$$D_c = \sum_{j=1}^{c} f_{j(2)}(f_{j(2)} - 1)/\langle N(N-1)\rangle, \tag{4.109}$$

$$D_m = \max(D_r, D_c). \tag{4.110}$$

$$h_{ij} = f_{i(1)} f_{j(2)}/N, \tag{4.111}$$

$$g_{ij} = \mathrm{entier}\left(h_{ij}\right), \tag{4.112}$$

$$c_{ij} = g_{ij}\left(h_{ij} - \frac{1}{2}g_{ij} - \frac{1}{2}\right), \tag{4.113}$$

$$D_e = 2\sum_{i=1}^{r} \sum_{j=1}^{c} c_{ij}/\langle N(N-1)\rangle. \tag{4.114}$$

The correction in D_2 concerns the values that are expected for the dot-product under the null hypothesis of independence given the marginal distributions by the two raters. The index has been extended to several other research situations as discussed before (Popping 1985).

In the situation where the dictionary was developed during the coding process, the raters might end with different categories. It is the task of the investigator to decrease these to only one set of categories. Popping (1992) has proposed an algorithm to realize this; this algorithm is based on a discussion between the raters and eventually the investigator.

Once the categories are (in principle) definite, low agreement between the raters is still possible. Such low agreement is a consequence of the raters being not well enough trained and the categories being not sufficiently distinction able. If this omission turns out in the pilot phase of the investigation, adapt ions are still possible.

For the demonstration of a discussion between raters, two persons were given an open question and 75 answers on that question. They were asked to formulate categories and to assign the answers to these categories. The question was taken from an investigation on the opinion about going into pension. It asks: ‘What does your firm do for people who go into pension shortly?’ The raters, who are denoted as *A* and *B*, have received information about this investigation and have operated independently. They have formulated the categories *a*1 to *a*4, respectively, *b*1 to *b*3. The descriptions of the categories follow below.

	rater A		*rater B*
*a*1	present and/or reception	*b*1	short time working
*a*2	shorter and less heavy work	*b*2	Souvenir
*a*3	pension fund	*b*3	Other
*a*4	advise and/or help		

In Table 4.21, the assignments of the answers are compared. The amount of agreement between the two raters expressed in the index *D*2 is 0.75 = (0.31 − 0.1200)/(0.3815 − 0.1200).

Next both raters have been invited to discuss their categories and assignments. If possible, they should change assignments or category labels in such a way that one set of categories results that is satisfying according to both. At a certain point in this discussion, it was decided to assign the answers again, but this time to reformulated categories. These categories are used by both raters and are as follows:

*c*1 short time work;
*c*2 present/reception;
*c*3 pension fund;
*c*4 information (pension);
*c*5 information (course);
*c*6 information (not specified);
*c*7 other.

The comparison of the assignments to the new categories resulted in *D*2 is 0.95. The raters agreed that ‘pension fund’ and ‘information about pension fund’ form in fact one category. Later they have joined these two categories. They did not want to

Table 4.21 Assignments by raters *A* and *B* compared

	Rater B	*b*1	*b*2	*b*3	#
Rater A	*a*1	1	35	0	36
	*a*2	10	0	0	10
	*a*3	0	0	16	16
	*a*4	0	0	10	10
	#	11	35	26	72

join the categories 'information (course)' and 'information (not specified),' because it is not ascertained that these categories refer to the same content. Besides it remains still possible to join them. Rater B has replaced the label 'souvenir' without discussion by 'present/reception.'

At this moment, the raters still disagreed about the assigning of three answers. These three answers are:

- They try to slow me down in my work;
- I will switch to less severe work;
- In the period until my pension, I will get an office job, so I do not have to go outside any longer.

After some discussion, in which also the frequency of the answers played a role, the raters agreed in joining the categories 'short time work' and 'other' to a category 'lighten of work.' At this moment, they agreed about the set of categories (that contains five categories) and the assignments of the answers to these categories.

Two decisions are very interesting in this discussion. The first one is the decision to place 'information about pension' in the pension fund category. The second one is that the raters have extended the number of categories, although partly because of pragmatic reasons. The discussion does not show real problems. Nowhere the raters disagreed categorical. Would this have been the case—no matter whether it concerns the formulation of categories or the assigning of the answers—then the project leader has to take the final decision.

When raters have to develop the categories by themselves, the sets of categories they end up with can be quite different. Popping (1983b) describes an experiment that investigates whether there are certain factors that can contribute to a higher level of agreement among the raters when this method of coding is used. In the experiment, answers to an open-ended question had to be coded. The factors that were examined include:

- Restricting the raters to a maximum number of categories;
- Presenting some information to the raters about the research project from which the question is taken and about the function of that question;
- Whether there is an optimal moment in the coding task at which this information can be provided; and
- Presenting the raters a set of categories developed by another rater who has performed the codings before under the same conditions.

Eighteen possible combinations of these factors were investigated for six questions posed to subjects whose responses constitute the observations to be classified, and that refer to attitudes, behavior, and factual information.

A major overall conclusion of the study is that agreement is low, even when the a posteriori method of coding is used. Differences in assignment of observations to categories contribute more to disagreement than differences in sets of categories. Another main conclusion was that the question itself is of prime importance. In cases where the question gives rise to a particular set of categories rather naturally, the amount of agreement is relatively high. In other cases, agreement is low because different sets of categories have been used in addition to any differences in assigning. The number of categories that is used also contributes to differences, especially in interaction with the question. However, no regular pattern was found in the kind of questions, the number of categories, and the interaction between these two factors.

It was proposed that the a priori method of coding is only used in pilot studies, for raters will differ substantially from each other in the sets of categories they initially formulate. After raters have finished their task, they should compare and discuss their classifications. All answers should then be reassigned to the categories that result from this discussion, and the discussion process should continue until a final set of categories results on which there is agreement. This set of categories should then be used in the main study.

4.3 Miscellaneous

4.3.1 Inference: Standard Errors, Significance, Interval Estimation

In a way it is questionable whether sampling characteristics for an agreement index are relevant. In the words of Cohen (1960: 44): '… it is generally of little value to test κ for significance… to know merely that κ is beyond chance is trivial since one usually expects much more than this in the way of reliability in psychological measurement. It may, however, serve as a minimum demand in some applications.'

For those who nevertheless want to compute the sampling characteristics, Mokken mentions a number of conditions that concern the values of coefficients in the population. Once the value of the coefficient in the sample has been determined, then there are the following problems of statistical analysis (Mokken 1971: 60):

1. Estimating the population coefficient from a random sample of N units;
2. Finding confidence intervals for this estimate;
3. Testing the null hypothesis that scores have been assigned on the basis of chance;
4. Testing the equality of coefficients for different populations.

Galtung (1967: 209) also highlights this point: The sampling distribution of the index must be known. This is especially relevant considering the generalizable reliability of the results. Galtung points out that finding this distribution is often very difficult.

Most interrater agreement studies using nominal categories suggest the existence of two populations of inference: the population of units and that of raters. Therefore, the sampling variance of the agreement coefficient can be considered as a result of the combined effect of both the sampling of units and of raters. The first would refer to the generalization of units and the second to the generalization of raters. The estimators for the variance, however, found in the literature account for the variability in the sampling of units. The extra sampling variance due to the sampling of raters is ignored. This probably is the part that contributes most to the variance.

When randomly selected units have been judged by a number of fixed raters, the resulting γ_3 value is an estimate of the corresponding parameter γ_3 for the population of units and the raters under consideration. Inferences about the parameter γ_3 can be drawn by computing an approximate confidence interval on the basis of the statistic γ_3 and its standard error.

No matter how large the sample of units may be, it can be that all units are assigned to one and the same category by all raters. In this case $P_o = P_e = 1$, and so the statistic γ_3 cannot be computed. One might say $\gamma_3 = 0$ because $P_o = P_e$; one might also define $\gamma_3 = 1$ because $P_o = 1$. A practical solution is to consider the distribution of γ_3 given that at least two different categories have been used. The problem demands a thorough theoretical examination, which is not within the scope of this text. When the sample of units is large, the statistic γ_3 can be written as a smooth function of multinomial proportions, which approximately follows a normal distribution with mean γ_3 and standard deviation that is to be estimated using one of the methods that follow.

The indices that follow below also concern only the estimation of the variance for the units. A first text in which the estimation of the rater variance is discussed is Gwet (2008b). All indices that have been proposed are generally used for descriptive purposes. In pilot studies however, often having a small number of units, reliability analyses are often used with the objective to use the results for judging a larger group from which the units that have been selected to far are considered as an a-select sample. This is all about the generalization to other units. This gives rise to the question whether the index found in the sample differs from the one in the population.

We will discuss only some main problems. In the population, with size N^*, ρ_{ij} indicates the probability that a random unit has been assigned by two raters to category i resp. j with marginal probabilities ρ_i and ρ_j. The frequencies f_{ij} in an a-select sample of size N are, as long as the sample size is small in comparison with the size of the population, multinomial distributed. For $N^* \to \infty$ and $N \to \infty$ with N/N^* being small, this distribution can be approached using a multinormal distribution, under the condition that none of the probabilities ρ_{ij} is very small.

Our index for comparing assignments by two raters following Cohen is:

$$\kappa = \left(\sum_{i=1}^{c} f_{ii} - \sum_{i=1}^{c} f_{i(1)} f_{i(2)}/N\right) \bigg/ \left(N - \sum_{i=1}^{c} f_{i(1)} f_{j(2)}/N\right). \tag{4.115}$$

When sampling is repeated both the marginal frequencies and the frequencies on the diagonal of the agreement table are subject to fluctuations. This might be discounted for in the formulas (and this will make sampling theory very complex) of one can consider the probability distribution of the index only in the subset of samples that have the observed marginal distributions. This is not unreasonable, and also in the computation of chance expected agreement, these frequencies are considered as being fixed. Given this condition on the marginal, the distribution of f_{ij} is multivariate hypergeometric:

$$P\left(f_{ij} = N_{ij} | f_{i.(1)} = N_{i(1).}; f_{j(2)} = N_{j(2)}\right) = \prod_{i=1}^{c} \frac{N_{i.(1)}!}{\prod_{j=1}^{c} N_{ij}!} / \frac{N!}{\prod_{j=1}^{c} N_{j(2)}!}. \quad (4.116)$$

Under the relatively mild conditions on the number N_{ij}, we assume that Σf_{ii} will be more or less normal distributed and the other expressions in the formula for π are fixed. Now the sampling theory for π is far simpler. In case weighting is applied not only the f_{ii} play a role but also the f_{ij} with i unequal j, but this is no real problem.

The distribution of the frequencies f_{ij} given the free marginals $f_{i(1)}$ and $f_{j(2)}$ are multinomial. Here too, the limit distribution is multivariate normal. It is not evident in all publications which assumptions have been used. In both models (marginal distribution fixed resp. free), however, π is a function of the frequencies f_{ij}. In case determining the complete probability distribution of π is too complex or not required, it is possible to determine at least the variance. As usually now the estimated proportions are used to indicate the population values. Besides in the model with free marginal distributions, the delta method is used (Bishop et al. 1975: 493), a development of series of the observed frequencies around the estimated ones. This implies that the formulas for the variances to follow are in nearly all estimations. These are in general satisfying for a reasonable N. Hubert (1977) has shown that the impact of fixed and free marginal distributions on the estimations is minor.

We present the formulas for the estimations of the variance for the most relevant indices. The pi's or kappa's in independent samples from a population have a distribution with mean μ and standard deviation σ. The general variance of the index is the variance no matter the value of the index in the population. This variance, also indicated as variance for the 'non-null case,' is relevant for computing confidence intervals.

In most agreement studies, a classification by raters is only available for a limited set of units. It is frequently desirable to be able to generalize the conclusions to a larger set of units, either those for which only one classification is available, or those that will be classified in the future. Statistical methods allow doing this provided that the units on which agreement is calculated are a random sample from this larger population of units.

In testing the significance of the amount of observed agreement, the null hypothesis is that raters are independent, with their marginal assignment probabilities equal to the observed marginal proportions. If Cohen is followed this means: the marginal proportion for each of the raters. Following Scott, it is the marginal proportion over all raters.

For all indices discussed, sampling characteristics are available. These have been derived specifically for the index or are estimated by using jackknife or bootstrap methods (Efron 1981). Due to space available, these sampling characteristics are not discussed/presented.

The indices in independent samples from a population have a distribution with mean μ and standard deviation σ. The index's general variance is the variance of the index independent of its value in the population. This variance, the variance of the non-null case, is relevant for computing confidence intervals. The variance under the null model, the variance under the null case, is sometimes found by means of a particularization of the general variance by replacing the P_o part in the formula by the P_e part. This part is used for testing the null hypothesis that the index is zero. Here, it is assumed that the sampling index is normally distributed. For this test, one uses:

$$z = \gamma/\sigma_0. \tag{4.117}$$

Connected to the z-value is a certain p-value. This p-value tests whether the estimated index is not due to chance. It does not test the strength of agreement. Also, p-values and confidence intervals are sensitive to sample size, and with a large enough sample size, any positive outcome will be statistically significant.

The formulas for the first versions of the standard deviation (and variance) as proposed by both Scott and Cohen were quite simple. They were based on P_o versus not P_o. The first formula to compute the standard deviation was:

$$\sigma = \sqrt{\frac{P_o(1 - P_o)}{N(1 - P_e)^2}}. \tag{4.118}$$

Under the null model of independence:

$$\sigma_0 = \sqrt{\frac{P_e}{N(1 - P_e)}}. \tag{4.119}$$

Later more detailed versions of these indices became available. In these versions, several aspects have been incorporated, which are not in the indices above: the sample size, whether weights are included. For the case of two fixed raters, Fleiss et al. 1969) derived the approximate large sample variances of kappa as defined by Cohen. For the case of more than two fixed raters, Davies and Fleiss (1982) presented formulas to estimate the variance under independence if there are only two categories. Schouten (1982b) presented formulas to estimate the variance of kappa under dependence and independence. In case the raters are varying, but with a constant number of raters per unit, so when Scott is followed, Fleiss et al. (1979) derived the estimated variance of pi under independence, and Schouten (1982a) derived the estimated variance of pi under dependence and under independence. For the situation of two categories, Fleiss and Cuzick (1979) derived formulas to estimate the variance of pi under independence when the number of raters is not the same for all units. Schouten (1986) developed these estimates for the situation of more than two categories.

Sampling characteristics for the index to be used in case a priori coding is applied are also known (Popping 1985).

An easy way to compute the variance of any of the indices that has been discussed is to use a jackknife estimation (Efron 1981; Hervé and Williams 2010). The goal of the jackknife is to estimate a parameter of a population of interest from a random sample of data from this population.

Let J be the statistic (such as γ_3) computed from the complete sample of N units and let $J_{\_s}$ be the value of the statistic when the sth unit is deleted from the sample. The sth pseudo-value is computed as

$$J_s = NJ - (N-1)\,J_{\_s}\,. \tag{4.120}$$

The Jackknife estimate of the index now is the average of these pseudo-values:

$$J^* = \sum_{s=1}^{N} J_s/N. \tag{4.121}$$

The entity

$$N(J^* - J)^{1/2}/\sigma_{J^*} \tag{4.122}$$

is under the hypothesis $J = J^*$ about Student t distributed with N − 1 degrees of freedom. Although this estimate has a smaller bias than J. In most applications, however, the difference between J^* and J will be negligible compared to their standard error. In large samples of units, an estimate of the variance of J^* and J may be taken as

$$\sigma_{J^*}^2 = \sigma_J^2 = \sum_{s=1}^{N} (J_s - J)^2/\langle N(N-1)\rangle. \tag{4.123}$$

The lower border for the confidence interval for the index now is:

$$J - t_{N-1}\sigma_{J^*}/N^{1/2}. \tag{4.124}$$

With respect to the distribution of the sampling kappa according to Cohen under the null model, Cicchetti and Fleiss (1977) found that for $3 \le c \le 7$ and Cicchetti (1981) that for $8 \le c \le 10$ the rule of thumb that the normality of this z-formula is very good, also in the situation of rather skewed marginals, under the condition that $N \ge 2c^2$. The situation with more than ten categories has not been investigated. Fleiss and Cicchetti (1978) have shown that one has to be careful in case weighted kappa for two raters is computed. In the situation that $3 \le c \le 5$ and

$0.4 \leq \kappa \leq 0.9$ one should be careful in setting up confidence intervals around κ when the sample size N is less than $16c^2$.

A final remark might be useful in this section. The sample used to compute agreement sometimes is a part of the complete sample used for the study that is going on. This complete sample often is not one drawn a select from a well-defined population. As it might be difficult to get data (e.g., when complex diseases are investigated), the investigator uses the units that satisfy the criteria for inclusion in the investigation, until a certain number is reached. It is hoped or assumed that this number is a good representation of the potential population.

4.3.2 Standards for Assessing and Reporting Reliability

Interrater agreement should be reported in a careful, clear, and detailed way in all research reports. Even if the assessment of this agreement is adequate, readers can only evaluate a study based on the information provided, which must be both complete and clear. Provide the following minimum information (based on Lombard et al. 2002: 602):

- The size of and the method used to create the agreement sample, along with a justification of that method;
- The relationship of the agreement sample to the full sample (i.e., whether the agreement sample is the same as the full sample, a subset of the full sample, or a separate sample);
- The number of agreement raters (which must be 2 or more) and whether or not they include the researcher(s);
- The amount of coding conducted by each agreement, and non-agreement, rater;
- The index or indices selected to calculate agreement and a justification of this/these selections;
- The interrater agreement level for each variable, for each index selected;
- The approximate amount of training (in hours) required to reach the agreement levels reported;
- How disagreements in the agreement coding were resolved in the full sample;
- Where and how the reader can obtain detailed information regarding the coding instrument, procedures, and instructions (e.g., from the authors).

I would like to add to this list the motivation for using the type of index that is computed. This implies most of all that the investigator has to report in more detail on the assignment procedure.

Usually during and immediately after training, the assignments by the raters are compared to those by an already trained rater. This is especially to see where differences are in assignments and why these differences exist. Some rater must have come with other interpretations than other raters.

The ‘official’ determination of the reliability is based on the coding of a part of the data set. Here, the investigator can choose between a representative part or a part containing all ‘problems’ the rater might run into.

4.3.3 *Computer Programs*

Computer programs are not listed here. Some indices, usually κ, are incorporated in commercial programs for statistical analysis as part of the indices that can be computed based on the data in a cross tabulation. Several individual developers, including myself, have developed specific programs to compute the agreement indices. These programs might suffer from running into incorrect outcomes or in not proceeding anymore, especially in the situation of extreme data. Users of the programs usually are not aware of potential problems with their data. Besides these developers created the program for their own purposes and that’s it. They will not take care of, or have no time for, adequate maintenance of the programs.

Most computer programs demand as input an agreement table in which the assignments by two raters are presented, or a data matrix like in Table 4.9. Here, the rows refer to the units, the columns to the raters, and the cell contains the number referring to the category the corresponding unit was assigned to by the rater in the column.

Some programs are stand-alone programs (like Popping 2010). Such programs demand that the data for which agreement is to be computed are kept in a separate file. The program by Hayes and Krippendorff (2007) is a macro in the statistical program SPSS. A series of routines in *R* has been provided by Gamer et al. (2010).

4.4 Indices that Do Not Fulfill the Requirements

Indices of type γ_3 fulfill the requirements that have been discussed before. Nevertheless, a lot of indices of type γ_0, γ_1, and γ_2 are available. Below these indices are listed, they are not discussed any further. The majority of these indices have only been defined for the situation in which assignments by two raters are compared. Besides many indices can only handle dichotomous data: Something applies or not. With respect to such data, it is impossible to compute the agreement per category. Several of the indices can easily be extended to an index of a higher type, but then they will often coincide with an index that was already defined by another author.

1. The maximum possible value of the index is 1, regardless of the number of raters or categories (max);
2. In the case of independence given the frequency in which each category is used by each rater or over all raters, the index takes the value 0 (ind);

3. Each rater agrees perfectly with himself (per);
4. Perfect agreement is a transitive relationship (trans);
5. Permutations of categories may not lead to other results. Because the data are measures on a nominal scale, the order of the categories is arbitrary and does not influence the results (permut);
6. The estimated value of the index is independent of the number of units (nobs);
7. If there are more than two categories, it should be possible to compute the amount of agreement for all categories together, but also per single category (cat);
8. If there are more than two raters, it should be possible to compute the amount of agreement for all raters together, but also per single rater (ram);
9. The index should be symmetrical with respect to the raters or ratings involved. An exception can be made for the situation in which one of the raters is taken as a standard (symm);
10. The sampling distribution of the index or at least the variance should be presented by the proposing author, at least approximately (var).

The requirements on robustness, interpretability, and validity are not mentioned. It cannot be proved whether they hold or not.

Below follow per type of index γ_0 to γ_3 the known indices together with an indication of whether the 10 requirements listed above apply or not by using a + or a − sign. First, the overview is presented for the situation of a priori coding, after that for the situation of a posteriori coding. Requirement 10 is easily solved by applying the jackknife or the bootstrap technique. Below only a + sign is used when a specific measure for the variance had been developed. The tables are for the greater part taken from Popping (1988).

Indices of type γ_0, show the number of units on which the raters agree, a posteriori coding (Table 4.22)

Indices of type γ_1, for comparing the assignments by two raters; SDAI is for comparing ratings (Table 4.23).

The indices *M* (Clement 1976) and *C* (House et al. 1981) have been developed for the situation in which one rater is a standard; MAI and MPDI are for comparing ratings. In case of perfect agreement, the index I (Woodward and Franzen 1948)

Table 4.22 Indices of type γ_0, count the number of units on which the raters agree, a posteriori coding

Requirement	1	2	3	4	5	6	7	8	9	10
	Max.	Ind.	Per.	Trans.	Permut.	Nobs.	Cat.	Ram.	Symm.	Var.
Index										
Y(1) (Elston et al. 1982)	−	−	−	+	+	+	*n*	+	+	+
SDAI (Armitage et al. 1966)	−	−	*n*	−	−	−	*n*	*n*	+	+

+ satisfies; − does not satisfy; *n* not appropriate

Table 4.23 Indices of type γ_1, the proportion of units on which the raters agree, range 0 to +1, a posteriori coding

Requirement	1	2	3	4	5	6	7	8	9	10
	Max.	Ind.	Per.	Trans.	Permut.	Nobs.	Cat.	Ram.	Symm.	Var.
Index										
Proportion observed agreement	+	−	+	+	+	+	+	+	+	+
II (Hawkins and Dotson 1975)	+	−	+	+	+	+	*n*	+	+	+
SI (Hawkins and Dotson 1975)	+	−	+	+	−	+	*n*	+	+	+
UI (Hawkins and Dotson 1975)	+	−	+	+	+	+	*n*	+	+	+
Mean SI and UI (Hawkins and Dotson 1975)	+	−	+	+	+	+	*n*	+	+	−
OI (Hopkins and Hermann 1977)	−	−	+	+	+	+	*n*	+	+	+
CR (Holsti 1969)	+	−	+	+	+	+	+	+	+	−
Voting concordance (Rae and Taylor 1970)	+	−	+	+	+	+	*n*	+	+	−
Solidarity (Rae and Taylor 1970)	+	+	+	+	−	+	*n*	+	+	−
M (Harris and Lahey 1978)	+	−	+	+	+	+	*n*	+	+	−
Yule's Q (Montgommery and Crittenden 1977)	+	−	+	−	+	+	*n*	+	+	−
Rand (Anderberg 1973)	+	−	+	−	+	−	*n*	+	+	−
DOT (Anderberg 1973)	−	−	+	−	+	−	*n*	+	+	−
Jaccard (Anderberg 1973)	+	−	+	−	+	−	*n*	+	+	−
SS1(Anderberg 1973)	+	−	+	−	+	−	*n*	+	+	−
S(d) (Dice 1945)	+	−	+	+	−	+	*n*	+	+	−
A2 (Rogot and Goldberg 1966)	+	−	+	+	+	+	*n*	+	+	−
A1 (Rogot and Goldberg 1966)	+	−	+	+	+	+	*n*	+	+	−
I (Woodward and Franzen 1948)	−	−	+	+	+	+	+	+	+	−
Y(3) (Elston et al. 1982)	+	−	−	+	+	+	*n*	+	+	+
A (Cartwright 1956)	+	−	+	+	+	+	+	*n*	+	+
MAI (Armitage et al. 1966)	−	−	*n*	+	+	+	*n*	*n*	+	+
MPDI (Armitage et al. 1966)	−	−	*n*	−	−	−	*n*	*n*	+	+
M (Clement 1976)	+	−	*n*	*n*	+	+	*n*	*n*	*n*	−
C (House et al. 1981)	−	−	*n*	−	−	−	*n*	*n*	*n*	−

+ satisfies; − does not satisfy; *n* not appropriate

Table 4.24 Indices of type γ_2, the proportion of units on which the raters agree, range -1 to +1, a posteriori coding

Requirement	1	2	3	4	5	6	7	8	9	10
	Max.	Ind.	Per.	Trans.	Permut.	Nobs.	Cat.	Ram.	Symm.	Var.
Index										
G (Holley and Guilford 1964)	+	−	+	+	+	+	*n*	+	+	+
RE (Maxwell 1977)	+	−	+	+	+	+	*n*	+	+	+
RE (Janes 1979)	+	−	+	+	+	+	+	+	+	+
R* (Fleiss 1965)	+	−	+	+	+	+	*n*	−	+	−
α (Cartwright 1956)	+	−	+	+	+	+	+	*n*	+	+
I_r (Perreault and Leigh 1989)										

+ satisfies; − does not satisfy; n not appropriate

Table 4.25 - Indices of type γ_3, the proportion of units on which the raters agree corrected for chance, range −1 to +1, a posteriori coding

Requirement	1	2	3	4	5	6	7	8	9	10
	Max.	Ind.	Per.	Trans.	Permut.	Nobs.	Cat.	Ram.	Symm.	Var.
Index										
S (Bennett et al. 1954)	+	−	+	+	+	+	+	+	+	+
α (Hayes and Krippendorff 2007)	+	+	+	+	+	+	+	+	+	−
AC1 (Gwet 2008a)	+	+	+	+	−	+	*n*	+	+	−
π (Flanders 1967)	+	−	+	+	+	+	+	+	+	−
π (Garrett 1975)	+	−	+	+	+	+	+	+	+	−
λ (Goodman and Kruskal 1954)	+	−	+	+	+	+	+	+	+	−
R11 (Maxwell and Pilliner 1968)	+	+	+	+	+	+	*n*	+	+	−
S (Janson and Vegelius 1979)	+	+	+	+	+	+	+	+	+	−
Occurrence κ (Kent and Foster 1977)	−	−	+	+	−	+	*n*	+	+	−
Non-occurrence κ (Kent and Foster 1977)	−	−	+	+	−	+	*n*	+	+	−
C (Cicchetti 1972)	−	+	+	+	+	+	+	+	+	+

+ satisfies; − does not satisfy; *n* not appropriate

Table 4.26 Indices of type γ_0, counts the number of pairs of units on which the raters agree, a priori coding

Requirement	1	2	3	4	5	6	7	8	9	10
	Max.	Ind.	Per.	Trans.	Permut.	Nobs.	Cat.	Ram.	Symm.	Var.
Index										
A' (Brennan and Light 1974)	−	−	−	+	+	−	−	−	+	+

+ satisfies; − does not satisfy; *n* not appropriate

Table 4.27 - Indices of type γ_1, the proportion of pairs of units on which the raters agree, range 0 to +1, a priori coding

Requirement	1	2	3	4	5	6	7	8	9	10
	Max.	Ind.	Per.	Trans.	Permut.	Nobs.	Cat.	Ram.	Symm.	Var.
Index										
Coefficient by Kulczynski (Anderberg 1973)	+	−	+	−	+	−	−	+	+	−
Coefficient by Sokal and Sneath SS2 (Anderberg 1973)	+	−	+	−	+	−	−	+	+	−

+ satisfies; − does not satisfy; *n* not appropriate

takes the value zero. Several of the indices above were developed as similarity indices to be used in cluster analysis (Anderberg 1973). Besides some indices have been developed in a specific field but are identical to similarity measures. In this way, interval by interval (II) is just the proportion on the diagonal, so it is equal to the Rand coefficient: $(f_{11} + f_{00})/N$; scored interval (SI) is identical to the coefficient of Jaccard. (Tables 4.24 and 4.25)

Note, both π and κ would have been included in the list above in case one of the requirements was not fulfilled. In both indices, chance expected agreement is used, but it is defined in a different way. The index α also fulfills the requirements, but the authors did not derive the variance. With respect to this index, chance expected agreement is computed in still another way than in π and κ (Table 4.26).

Table 4.28 Indices of type γ_2, the proportion of pairs of units on which the raters agree, range −1 to +1, a priori coding

Requirement	1	2	3	4	5	6	7	8	9	10
	Max.	Ind.	Per.	Trans.	Permut.	Nobs.	Cat.	Ram.	Symm.	Var.
Index										
Γ (Hubert 1977)	+	−	−	+	+	+	−	−	+	+
J (Janson and Vegelius 1982)	+	+	+	+	+	+	−	+	+	+
Product moment correlation	+	−	+	−	+	+	−	+	+	−

+ satisfies; − does not satisfy; *n* not appropriate

Indices of type γ_1, the proportion of pairs of units on which the raters agree, range 0 to +1, a priori coding (Table 4.27).

Indices that have been derived for comparing similarity between assignments in cluster analysis can also be used here. The indices will not be based on the number of units, but on the number of pairs of units.

Type γ_2, outcomes in range −1 to +1, a priori coding (Table 4.28).

The only known index of type γ_3 which corrects for chance expected agreement has outcomes in the range −1 to +1 and is to be used when a posteriori coding is applied is the index *D*2, which was discussed before.

References

A. Agresti, Modelling patterns of agreement and disagreement. Stat. Methods Med. Res. **1**(2), 201–218 (1992)

M.R. Anderberg, *Cluster Analysis for Applications* (Academic Press, New York, 1973)

P. Armitage, L.M. Blendis, H.C. Smyllie, The measurement of observer disagreement in the recording of signs. J. R. Stat. Soc. (A) **129**(1), 98–109 (1966)

W. Barlow, M.Y. Lai, S.P. Azen, A comparison of methods for calculating a stratified kappa. Stat. Med. **10**(9), 1465–1472 (1991)

E.M. Bennett, R.L. Blomquist, A.C. Goldstein, Response stability in limited response questioning. Public Opin. Q. **18**(2), 218–223 (1954)

Y.M.M. Bishop, S.E. Fienberg, P.W. Holland, *Discrete Multivariate Analysis. Theory and Practice* (The MIT Press, Cambridge, 1975)

R.L. Brennan, R.J. Light, Measuring agreement when two observers classify people into categories not defined in advance. Br. J. Math. Stat. Psychol. **27**(2), 154–163 (1974)

T. Byrt, J. Bishop, B. Carlin, Bias, prevalence and kappa. J. Clin. Epidemiol. **46**(5), 423–429 (1993)

D.S. Cartwright, A rapid non parametric estimate of multi judge reliability. Psychometrika **21**(1), 17–29 (1956)

D.V. Cicchetti, A new measure of agreement between rank ordered variables, in *Proceedings of the 80th Annual Convention*, vol. 7. American Statistical Association (1972), pp. 17–18

D.V. Cicchetti, Testing the normal approximation and minimal sample requirements of weighted kappa when the number of categories is large. Appl. Psychol. Meas. **5**(1), 101–104 (1981)

D.V. Cicchetti, J.L. Fleiss, Comparing the null distributions of weighted kappa and the C ordinal statistic. Appl. Psychol. Meas. **1**(2), 195–201 (1977)

D.V. Cicchetti, C. Lee, A.F. Fontana, B.N. Dows, A computer program for assessing specific category rater agreement for qualitative data. Educ. Psychol. Measur. **38**(3), 805–813 (1978)

P.G. Clement, A formula for computing inter-observer agreement. Psychol. Rep. **39**(1), 257–258 (1976)

J. Cohen, A coefficient of agreement for nominal scales. Educ. Psychol. Measur. **20**(1), 37–46 (1960)

A.J. Conger, Integration and generalization of kappas for multiple raters. Psychol. Bull. **88**(2), 322–328 (1980)

R.T. Craig, Generalization of Scott's index of intercoder agreement. Public Opinion Quarterly **45** (2), 260–264 (1981)

M. Davies, J.L. Fleiss, Measuring agreement for multinomial data. Biometrics **38**(4), 1047–1051 (1982)

L.R. Dice, Measures of the amount of ecologic association between species. Ecology **26**(3), 297–302 (1945)

B. Efron, Nonparametric estimates of standard error: The jackknife, the bootstrap and other methods. Biometrika **68**(3), 589–599 (1981)

R.C. Elston, S.R. Schroeder, J. Rojahn, Measures of observer agreement when binomial data are collected in free operant situations. J. Behav. Assess. **4**(4), 299–310 (1982)

A.R. Feinstein, D.V. Cicchetti, High agreement but low kappa, I: the problems of two paradoxes. J. Clin. Epidemiol. **43**(6), 543–549 (1990)

N.A. Flanders, *Interaction Analysis: Theory, Research and Applications* (Addison-Wesley, Reading, 1967), pp. 161–166

J.L. Fleiss, Estimating the accuracy of dichotomous judgements. Psychometrika **30**(4), 469–479 (1965)

J.L. Fleiss, Measuring nominal scale agreement among many raters. Psychol. Bull. **76**(5), 378–382 (1971)

J.L. Fleiss, D.V. Cicchetti, Inference about weighted kappa in the non-null case. Appl. Psychol. Meas. **2**(1), 113–117 (1978)

J.L. Fleiss, J. Cohen, B.S. Everitt, Large sample standard errors of kappa and weighted kappa. Psychol. Bull. **72**(5), 323–327 (1969)

J.L. Fleiss, J. Cuzick, The reliability of dichotomous judgments: unequal number of judges per subject. Appl. Psychol. Meas. **3**(4), 537–542 (1979)

J.L. Fleiss, B.S. Everitt, Comparing the marginal totals of square contingency tables. Br. J. Math. Stat. Psychol. **24**(1), 117–123 (1971)

J.L. Fleiss, J.C.M. Nee, J.R. Landis, Large sample variance of kappa in the case of different sets of raters. Psychol. Bull. **86**(5), 974–977 (1979)

J.L. Fleiss, R.L. Spitzer, J. Endicott, J. Cohen, 'Quantification of agreement in multiple psychiatric diagnosis. Arch. Gen. Psychiatry **26**(2), 168–171 (1972)

J.L. Fleiss, *Statistical Methods for Rates and Proportions* (Wiley, New York, 1981)

J. Galtung, *Theory and Methods of Social Research* (Allen & Unwin, London, 1967)

M. Gamer, J. Lemon, I. Fellows, P. Sing, P, in *Various coefficients of interrater reliability and agreement*. (Version 0.83) [software] (2010). Available from http://CRAN.R-project.org/package=irr

C.S. Garrett, Modification of the Scott coefficient as an observer agreement estimate for marginal-form observation scale data. J Exp Educ **43**(1), 21–26 (1975)

L.A. Goodman, W.H. Kruskal, Measures of association for cross classifications. J. Am. Stat. Assoc. **49**(268), 732–764 (1954)

W.M. Grove, N.C. Andreasen, P. McDonald-Scott, M.B. Keller, R.W. Shapiro, Reliability studies of psychiatric diagnosis. Theory and practice. Arch. Gen. Psychiatry **38**(4), 408–413 (1981)

K.L. Gwet, Computing inter-rater reliability and its variance in the presence of high agreement. Br. J. Math. Stat. Psychol. **61**(1), 29–48 (2008a)

K.L. Gwet, Variance estimation of nominal-scale inter-rater reliability with random selection of raters. Psychometrika **73**(3), 407–430 (2008b)

F.C. Harris, B.B. Lahey, A method for combining occurrence and nonoccurrence interobserver agreement scores. J. Appl. Behav. Anal. **11**(4), 523–527 (1978)

R.P. Hawkins, V.A. Dotson, Reliability scores that delude: an Alice in Wonderland trip through the misleading characteristics of interobserver agreement scores in interval recording, in *Behavior Analysis. Areas of Research and Application*, ed. by E. Ramp, G. Semb (Prentice Hall, Englewood Cliffs, 1975), pp. 359–376

A.F. Hayes, K. Krippendorff, Answering the call for a standard reliability measure for coding data. Commun. Methods Measures **1**(1), 77–89 (2007)

A. Hervé, L.J. Williams, Jackknife, in *Encyclopedia of Research Design*, ed. by N. Salkind (Sage, Thousand Oaks, 2010)

J.W. Holley, J.P. Guilford, A note on the G-index of agreement. Educ. Psychol. Measur. **24**(4), 749–753 (1964)

O.R. Holsti, *Content Analysis for the Social Sciences and Humanities* (Addison Wesley, London, 1969)

B.L. Hopkins, J.A. Hermann, Evaluating interobserver reliability of interval data. J. Appl. Behav. Anal. **10**(1), 121–126 (1977)

A.E. House, B.J. House, M.B. Campbell, Measures of interobserver agreement: calculation formulas and distribution effects. J. Behav. Assess. **3**(1), 37–57 (1981)

L.J. Hubert, Kappa revisited. Psychol. Bull. **84**(2), 289–297 (1977)

C.L. Janes, Extension of the random error coefficient of agreement to N × N tables. Br. J. Psych. **134**(6), 617–619 (1979)

S. Janson, J. Vegelius, On the generalization of the G-index and the phi coefficient to nominal scales. Multivar. Behav. Res. **14**(2), 255–269 (1979)

S. Janson, J. Vegelius, The J-index as a measure of nominal scale response agreement. Appl. Psychol. Measure. **6**(1), 111–121 (1982)

R.N. Kent, S.L. Foster, Direct observational procedures: methodological issues in naturalistic settings, in *Handbook of Behavioral Assessment*, ed. by A.R. Ciminero, K.S. Calhoun, H.E. Adams (Wiley, New York, 1977), pp. 279–328

H.C. Kraemer, Extensions of the kappa coefficient. Biometrics **36**(2), 207–216 (1980)

L.L. Kupper, K.B. Hafner, On assessing interrater agreement for multiple attribute responses. Biometrics **45**(3), 957–967 (1989)

J.R. Landis, G.G. Koch, A review of statistical methods in the analysis of data arising from observer reliability studies. Part 2. Stat. Neerl. **29**(2), 151–161 (1975)

J.R. Landis, G.G. Koch, The measurement of observer agreement for categorical data. Biometrics **33**(1), 159–174 (1977)

J. Loevinger, The technique of homogeneous test compared with some aspects of 'scale analysis' and 'factor analysis'. Psychol. Bull. **45**(6), 507–530 (1948)

M. Lombard, J. Snyder-Duch, C.C. Bracken, Content analysis in mass communication: assessment and reporting of intercoder reliability. Human Commun. Res. **28**(4), 587–604 (2002)

A.E. Maxwell, Comparing the classifications of subjects by two independent judges. Br. J. Psych. **116**(535), 651–655 (1970)

A.E. Maxwell, Coefficients of agreement between observers and their interpretation. Br. J. Psych. **130**(1), 79–83 (1977)

A.E. Maxwell, A.E.G. Pilliner, Deriving coefficients of reliability and agreement for ratings. Br. J. Math. Stat. Psychol. **21**(1), 105–116 (1968)

R.J. Mokken, *A Theory and Procedure of Scale Analysis: With Applications in Political Research* (Mouton,The Hague, 1971)

A.C. Montgomery, K.S. Crittenden, Improving coding reliability for open-ended questions. Public Opin. Q. **41**(2), 235–243 (1977)

W.D. Perreault, L.E. Leigh, Reliability of nominal data based on qualitative judgments. J. Mark. Res. **26**(2), 135–148 (1989)

R. Popping, Traces of agreement. On the dot-product as a coefficient of agreement. Qual. Quant. **17**(1), 1–18 (1983)

R. Popping, Overeenstemmingsmaten voor nominale data [Measures of agreement for nominal data]. Unpublished PhD, University of Groningen, Groningen (1983b)

R. Popping, Traces of agreement: extensions of the D2-index. Qual. Quant. **19**(4), 383–388 (1985)

R. Popping, On agreement indices for nominal data, in *Sociometric Research*, vol. I, ed. by W.E. Saris, I.N. Gallhofer (McMillan, London, 1988), pp. 90–105

R. Popping, In search for one set of categories. Qual. Quant. **25**(1), 147–155 (1992)

R. Popping, Ag09. A computer program for interrater agreement for judgments. Social Sci. Comput. Rev. **28**(3), 391–396 (2010)

D.W. Rae, M. Taylor, *The Analysis of Political Cleavages* (Yale University Press, New Haven, 1970), pp. 115–145

E. Rogot, I.D. Goldberg, A proposed index for measuring agreement in test-retest studies. J. Chronic Diseases **19**(9), 991–1006 (1966)

D.C. Ross, Testing patterned hypothesis in multi-way contingency tables using weighted kappa and weighted chi square. Educ. Psychol. Measur. **37**(2), 291–307 (1977)

H.J.A. Schouten, Measuring pairwise agreement among many observers. Biometr. J. **22**(6), 497–504 (1980)

H.J.A. Schouten, Measuring pairwise agreement among many observers. II. Some improvements and additions. Biometr. J. **24**(5), 431–435 (1982a)

H.J.A. Schouten, Measuring pairwise interobserver agreement when all subjects are judged by the same observers. Stat. Neerl. **36**(2), 45–61 (1982b)

H.J.A. Schouten, Nominal scale agreement among observers. Psychometrika **51**(3), 453–466 (1986)

P.E. Shrout, R.L. Spitzer, J.L. Fleiss, Quantification of agreement in psychiatric diagnosis revisited. Arch. Gen. Psychiatry **44**(2), 172–177 (1987)

J. Spanjer, B. Krol, R. Popping, J.W. Groothoff, S. Brouwer, Disability assessment interview: the role of concrete and detailed information on functioning besides medical history taking. J. Rehabil. Med. **41**(4), 267–272 (2009)

A. Stuart, A test of homogeneity of marginal distributions in a two-way classification. Biometrika **42**(3/4), 412–416 (1955)

J.S. Uebersax, A generalized kappa coefficient. Educ. Psychol. Measu. **42**(1), 181–183(1982)

J.S. Uebersax, Diversity of decision-making models and the measurement of interrater agreement. Psychol. Bull. **101**(1), 140–146 (1987)

M.J. Warrens, Cohen's kappa can always be increased and decreased by combining categories. Stat. Method. **7**(6), 673–677 (2010)

G.W. Williams, Comparing the joint agreement of several raters with another rater. Biometrics **32** (3), 619–627 (1976)

J.L. Woodward, R. Franzen, A study on coding reliability. Public Opin. Q. **12**(2), 253–257 (1948)

R. Zwick, Another look at interrater agreement. Psychol. Bull. **103**(3), 374–378 (1988)

Index

R. Popping, *Introduction to Interrater Agreement for Nominal Data*,
https://doi.org/10.1007/978-3-030-11671-2

MIX
Papier aus verantwortungsvollen Quellen
Paper from responsible sources
FSC® C105338

If you have any concerns about our products,
you can contact us on
ProductSafety@springernature.com

In case Publisher is established outside the EU,
the EU authorized representative is:
Springer Nature Customer Service Center GmbH
Europaplatz 3, 69115 Heidelberg, Germany

Printed by Libri Plureos GmbH
in Hamburg, Germany